COMMUNITY COLLEGES
IN THE EVOLVING
STEM EDUCATION LANDSCAPE

Summary of a Summit

Steve Olson and Jay B. Labov, *Rapporteurs*

Planning Committee on Evolving Relationships and Dynamics Between
Two- and Four-Year Colleges and Universities

Board on Higher Education and Workforce
Division on Policy and Global Affairs

Board on Life Sciences
Division on Earth and Life Studies

Board on Science Education
Teacher Advisory Council
Division of Behavioral and Social Sciences and Education

Engineering Education Program Office
National Academy of Engineering

NATIONAL RESEARCH COUNCIL *AND*
NATIONAL ACADEMY OF ENGINEERING
OF THE NATIONAL ACADEMIES

THE NATIONAL ACADEMIES PRESS
Washington, D.C.
www.nap.edu

THE NATIONAL ACADEMIES PRESS   500 Fifth Street, NW   Washington, DC 20001

NOTICE: The project that is the subject of this report was approved by the Governing Board of the National Research Council, whose members are drawn from the councils of the National Academy of Sciences, the National Academy of Engineering, and the Institute of Medicine. The members of the committee responsible for the report were chosen for their special competences and with regard for appropriate balance.

This study was supported by Grant No. EHR 1112988 between the National Academy of Sciences and the National Science Foundation, and in-kind support from the Carnegie Institution for Science. Any opinions, findings, conclusions, or recommendations expressed in this publication are those of the author and do not necessarily reflect the views of the organizations or agencies that provided support for the project.

International Standard Book Number-13:   978-0-309-25654-4
International Standard Book Number-10:   0-309-25654-2

Additional copies of this report are available from the National Academies Press, 500 Fifth Street, NW, Keck 360, Washington, DC 20001; (800) 624-6242 or (202) 334-3313; http://www.nap.edu/.

Suggested citation: National Research Council and National Academy of Engineering. (2012). *Community Colleges in the Evolving STEM Education Landscape: Summary of a Summit.* S. Olson and J.B. Labov, Rapporteurs. Planning Committee on Evolving Relationships and Dynamics Between Two- and Four-Year Colleges, and Universities. Board on Higher Education and Workforce, Division on Policy and Global Affairs. Board on Life Sciences, Division on Earth and Life Studies. Board on Science Education, Teacher Advisory Council, Division of Behavioral and Social Sciences and Education. Engineering Education Program Office, National Academy of Engineering. Washington, DC: The National Academies Press.

THE NATIONAL ACADEMIES
Advisers to the Nation on Science, Engineering, and Medicine

The **National Academy of Sciences** is a private, nonprofit, self-perpetuating society of distinguished scholars engaged in scientific and engineering research, dedicated to the furtherance of science and technology and to their use for the general welfare. Upon the authority of the charter granted to it by the Congress in 1863, the Academy has a mandate that requires it to advise the federal government on scientific and technical matters. Dr. Ralph J. Cicerone is president of the National Academy of Sciences.

The **National Academy of Engineering** was established in 1964, under the charter of the National Academy of Sciences, as a parallel organization of outstanding engineers. It is autonomous in its administration and in the selection of its members, sharing with the National Academy of Sciences the responsibility for advising the federal government. The National Academy of Engineering also sponsors engineering programs aimed at meeting national needs, encourages education and research, and recognizes the superior achievements of engineers. Dr. Charles M. Vest is president of the National Academy of Engineering.

The **Institute of Medicine** was established in 1970 by the National Academy of Sciences to secure the services of eminent members of appropriate professions in the examination of policy matters pertaining to the health of the public. The Institute acts under the responsibility given to the National Academy of Sciences by its congressional charter to be an adviser to the federal government and, upon its own initiative, to identify issues of medical care, research, and education. Dr. Harvey V. Fineberg is president of the Institute of Medicine.

The **National Research Council** was organized by the National Academy of Sciences in 1916 to associate the broad community of science and technology with the Academy's purposes of furthering knowledge and advising the federal government. Functioning in accordance with general policies determined by the Academy, the Council has become the principal operating agency of both the National Academy of Sciences and the National Academy of Engineering in providing services to the government, the public, and the scientific and engineering communities. The Council is administered jointly by both Academies and the Institute of Medicine. Dr. Ralph J. Cicerone and Dr. Charles M. Vest are chair and vice chair, respectively, of the National Research Council.

www.national-academies.org

PLANNING COMMITTEE ON EVOLVING RELATIONSHIPS AND DYNAMICS BETWEEN TWO- AND FOUR- YEAR COLLEGES AND UNIVERSITIES

GEORGE R. BOGGS (*Chair*), President Emeritus, American Association of Community Colleges
THOMAS R. BAILEY, Columbia University
LINNEA FLETCHER, Austin Community College
BRIDGET TERRY LONG, Harvard University
JUDY C. MINER, Foothill Community College
KARL S. PISTER, * University of California

JAY B. LABOV, *Senior Advisor for Education and Communication, Director, National Academies Teacher Advisory Council,* and *Project Study Director*
CATHERINE DIDION, *Senior Program Officer,* National Academy of Engineering, and *Project Co-Director*
PETER H. HENDERSON, *Director,* Board on Higher Education and Workforce, and *Project Co-Director*
MARGARET L. HILTON, *Senior Program Officer,* Board on Science Education
MARTIN STORKSDIECK, *Director,* Board on Science Education, and *Project Co-Director*

CYNTHIA A. WEI, *Christine Mirzayan Policy Fellow,* National Academy of Sciences (through December 16, 2012)
ORIN E. LUKE, *Senior Program Assistant*
MARY ANN KASPER, *Senior Program Assistant*

---

*Member, National Academy of Engineering

Acknowledgments

This summary has been reviewed in draft form by individuals chosen for their diverse perspectives and technical expertise, in accordance with procedures approved by the National Research Council's (NRC) Report Review Committee. The purpose of this independent review is to provide candid and critical comments that will assist the institution in making its published summary as sound as possible and to ensure that the summary meets institutional standards for objectivity, evidence, and responsiveness to the study charge. The reviewers' comments and draft manuscript remain confidential to protect the integrity of the process. We thank the following individuals for their review of this summary: Ashok Agrawal, vice president for academic affairs, and Department of Mechanical Engineering, St. Louis Community College at Florissant Valley; Cathleen Aubin Barton, education manager, Intel Corporation, Chandler, AZ; George R. Boggs, superintendent/president emeritus, Palomar College and president and CEO emeritus, American Association of Community Colleges; and Ronald Williams, vice president, The College Board, Washington, DC.

Although the reviewers listed above provided many constructive comments and suggestions, they were not asked to endorse the content of the report, nor did they see the final draft of the report before its release. The review of this report was overseen by Melvin D. George, president emeritus, St. Olaf College, and the University of Missouri System. Appointed by the NRC, he was responsible for making certain that an independent examination of this report was carried out in accordance with institutional procedures and that all review comments were carefully considered. Responsibility for the final content of this report rests entirely with the author and the institution.

We thank Orin Luke, senior program assistant, for his valuable contributions to planning and implementing the logistics for all aspects of the convocation. We also thank Cynthia Wei, former Christine Mirzayan Policy fellow of the National Academy of Sciences, and Rebecca Fischler, communication officer in the NRC's Division on Earth and Life Studies, for her critical expert advice and assistance with developing and maintaining the convocation's website (see http://nas-sites.org/community collegessummit/) and electronic procedures.

Special thanks are extended to Toby Horn, Carnegie Institution for Science, and the Carnegie Institution itself for working with the staff to provide the venue for this event and offering in-kind support to make the venue available to the committee and participants.

Finally, we thank all of the participants for taking the time and, for many, the expense to come to this convocation.

George Boggs, Ph.D.   Jay B. Labov, Ph.D.
Chair, Organizing Committee   *Study Director* and *Rapporteur*

Contents

1   Introduction   1

2   Expanding Minority Participation in Undergraduate STEM
    Education   11

3   The Loss of Students from STEM Majors   19

4   Outreach, Recruitment, and Mentoring   23

5   The Two-Year Curriculum in Mathematics   29

6   Transfer from Community Colleges to Four-Year Institutions   35

7   General Discussion   41

References   51

Appendixes

A   Summit Agenda   53
B   Effective Outreach, Recruitment, and Mentoring into
    STEM Pathways: Strengthening Partnerships with
    Community Colleges   57
       Becky Wai-Ling Packard

C   Two-Year College Mathematics and Student Progression in
     STEM Programs of Study   81
       Debra D. Bragg
D   Developing Supportive STEM Community College to
     Four-Year College and University Transfer Ecosystems   107
       Alicia C. Dowd
E   Brief Biographies of Committee Members and Staff   135
F   Brief Biographies of Presenters and Panelists   141

1

Introduction

Picture a first-generation new community college student named Josie, said Becky Wai-Ling Packard, professor of psychology and education at Mount Holyoke College, in her presentation at the summit Community Colleges in the Evolving STEM Education Landscape. Josie became interested in environmental issues in high school when she became aware of the extent to which they affected her community, so she decided to go to college to learn more about environmental policy.

First she had to fill out financial aid paperwork, but much of it was hard to decipher. She had to figure out which courses to take, but she was not sure about the difference between a one-year certificate and an associate's degree, and her parents were not able to help her. She had questions about how much time the degree was going to take her. Given her family's financial status, she could not afford any missteps.

Imagine, several years in the future, that Josie is one of the few first-generation college students and students of color who has earned an associate's degree in a science, technology, engineering, and mathematics (STEM) field and has transferred to a four-year university. At her new school, she felt lost once again. No advisor met with her. She did all of her scheduling online. Peer study groups had already formed. The hours of the academic resource center conflicted with her job. Faculty assumed that she was incompetent when her experiences did not align with their expectations. Summer research sounded like a critical experience, but it required 40 hours a week, and she had to work at a conventional job to earn money for school. Furthermore, as a transfer student, she did not know any professors well enough to ask for a letter of recommendation.

Many thousands of students are like Josie, said Packard—indeed, they are the most typical students in higher education today. The skills of the future workforce will depend to a critical degree on how well community colleges meet the needs of these students.

THE ROLE OF COMMUNITY COLLEGES[1]

Community colleges are an often overlooked but essential component in the U.S. STEM education system. About 1,200 community colleges in the United States enroll more than 8 million students annually, including 43 percent of U.S. undergraduates (American Association of Community Colleges, 2011; Mullin and Phillippe, 2011). Community colleges provide not only general education but also many of the essential technical skills on which economic development and innovation are based. Almost one-half of the Americans who receive bachelor's degrees in science and engineering attended community college at some point during their education, and almost one-third of recipients of science or engineering master's degree did so (Tsapogas, 2004). About 40 percent of the nation's teachers, including teachers of science and mathematics, completed some of their mathematics or science courses at community colleges (Shkodriani, 2004). Community colleges provide professional development programs for teachers, offer alternative teacher certification programs for people who have a degree in another field, and in some states award baccalaureate degrees in teacher education and other disciplines.

Community colleges provide the most diverse student body in the history of the United States with access to higher education. Community colleges serve people of color, women, older students, veterans, international students, first-generation college goers, and working parents. In particular, minorities who are underrepresented in STEM fields are disproportionately enrolled in community colleges. Fifty-two percent of Hispanic students, 44 percent of African American students, 55 percent of Native American students, and 45 percent of Asian-Pacific Islander students attend community colleges (American Association of Community Colleges, 2011).

---

[1]The remainder of this chapter is based on the introductory remarks made at the summit by George Boggs, president emeritus, American Association of Community Colleges; Barbara Olds, acting deputy director, Directorate for Education and Human Resources, National Science Foundation; Jane Oates, assistant secretary, Employment Training Administration, U.S. Department of Labor; and Toby Horn, co-director, Carnegie Academy for Science Education.

Community colleges are more affordable as well as more accessible than four-year institutions. Average tuition and fees at a community college are about $3,000 per year, compared with an average of $8,200 per year for in-state four-year institutions, $21,000 per year for out-of-state students at state institutions, and $29,000 per year at private institutions (College Board, 2011). Indeed, it is this large difference between the cost of attending community colleges versus even the least expensive four-year institutions, especially during difficult economic times, that serves as an impetus for many more students to begin their college careers at two-year institutions.

Community colleges also focus on teaching in an era when teaching in higher education is receiving particular scrutiny and calls for accountability. And community colleges are becoming an increasing focus of educational researchers as their contributions to education—and to STEM education in particular—are more widely recognized.

RATIONALE FOR THE SUMMIT

Given the increasing importance of community colleges in the U.S. STEM education system, the National Research Council of the National Academies and the Carnegie Academy for Science Education of the Carnegie Institution for Science hosted the Summit on Community Colleges in the Evolving STEM Education Landscape on December 15, 2011.[2] The event was hosted by the Carnegie Institution for Science in Washington, DC, with support from the National Science Foundation (NSF).

The importance of community colleges, especially in emerging areas of STEM and preparation of the STEM workforce, has been recognized for at least 20 years, e.g., through the establishment of the Advanced Technological Education Program at the National Science Foundation. However, given the attention that both community colleges and STEM education have received in recent years, combined with new ways of viewing the roles of community colleges in the nation's education system (e.g., dual enrollment for high school students, bi-directional pathways between community colleges and four-year institutions, and pre-service education for teachers), a thorough re-examination of the status, promise, and opportunities of community colleges and their contributions to STEM education is long overdue. Community college will be essential to accommodate growing numbers of students, especially given the Obama

---

[2]Planning for this summit was a collaborative effort of the Board on Higher Education and Workforce, the Board on Life Sciences, the Board on Science Education, and the Teacher Advisory Council of the National Research Council, and the Engineering Education Program Office of the National Academy of Engineering.

Administration's goals of increasing the number of college graduates by 5 million over 10 years, 3 million of whom would be educated by community colleges. Many have ongoing relationships with local community organizations, governments, and businesses that allow them to respond quickly to community needs. Community colleges retrain displaced workers in skills needed by local businesses and open gateways to individuals who would otherwise lack the preparation or financial resources to receive a college education. They prepare students for STEM occupations that require a certificate or associate's degree as well as for transfer to four-year institutions. They serve as models of excellence for STEM education in an increasingly global economy and in educating a highly prepared technical workforce.

To organize the summit, a planning committee was appointed by the chair of the National Research Council and charged with a Statement of Task (see Box 1-1).

The planning committee for the summit decided to focus the event on three critical areas:

1. Outreach and partnerships between community colleges and four-year institutions
2. Subjects that can serve as gateways or barriers to college completion, with college-level mathematics as an exemplar
3. Transfer issues for students from two-year to four-year colleges and universities

ORGANIZATION OF THIS REPORT

This report has been prepared by the workshop rapporteur as a factual summary of what occurred during the summit. Because the event was limited to one day, many topics that might have been discussed in the context of the summit's goals and objectives had to be omitted. However, the summit did raise a number of important additional questions (see reports of general discussions as well as comments from breakout sessions) that deserve additional attention in future convenings and as part of a future research agenda. In addition, the three commissioned papers that are provided in Appendixes B-D provide additional research evidence as well as questions and suggestions from the authors of those papers about a series of related questions and issues.

The planning committee's role was limited to planning and convening the workshop. The views contained in the report are those of individual workshop participants and do not necessarily represent the views of all workshop participants, the planning committee, or the National Research Council/Institute of Medicine.

BOX 1-1
Committee Statement of Task

An ad hoc committee will plan and conduct a summit that will feature invited presentations and discussions on science, technology, engineering, and mathematics (STEM) education in two-year higher education institutions and how the changing dynamics between two-year and four-year institutions of higher education might offer new educational opportunities for students, institutions, and the nation's workforce. The summit will include leaders from community colleges, four-year postsecondary institutions, business and industry, and state and federal policymakers, and researchers with expertise in community colleges, student learning, and teaching. The summit will allow these stakeholders to engage in discussions of critical issues in two-year and four-year STEM education. Experts will be commissioned to summarize the evidence available in several commissioned papers, which will be available prior to the planning meeting and summit. The presentations and discussions from the summit will be synthesized in an individually authored workshop summary.

The project will commence with a planning meeting in spring of 2011 to plan the summit and to discuss how program units of the National Academies, both individually and collectively, can best contribute to federal initiatives on community colleges in the future.

This activity will be conducted as a collaboration among the Board on Higher Education and Workforce, the Board on Life Sciences, the Board on Science Education, and the Teacher Advisory Council of the National Research Council, and the Engineering Education Program Office of the National Academy of Engineering.

Because many of the issues were discussed throughout the one-day summit, this summary provides a narrative rather than a chronological overview of the presentations and the rich discussions that permeated the event.

After this introductory chapter, Chapter 2 examines ways to expand the participation of underrepresented minorities in STEM fields, while Chapter 3 describes a study of why students who enter college intending to major in a STEM field switch to other majors.

As noted above, three commissioned papers were produced before the summit and posted on the summit's website prior to it.[3] The main points drawn from these papers and the discussions they provoked are discussed in Chapters 4 through 6 of this report, respectively.

Boxes at the beginning of Chapters 2-6 summarize the important points made in those chapters, and these boxes could be read as a quick

---

3The website is available at http://nas-sites.org/communitycollegessummit/.

Ongoing Initiatives

Many organizations have become interested in the role of community colleges in the U.S. education system, including the Business Higher Education Forum, the Association of American Universities, the Association of Public and Land Grant Universities, the Association of American Colleges and Universities, the Howard Hughes Medical Institute, the Lumina Foundation, the Gates Foundation, and the College Board, all of which sent representatives to the summit. In addition, numerous federal agencies, and divisions within federal agencies, are supporting initiatives in community colleges. For example, the Advanced Technology Education Program at NSF has made a strong and long-standing contribution to community colleges. Other programs at NSF, both in the Education and Human Resources directorate and in other directorates, also fund community college work, facilitate transitions for students among educational institutions, and support research involving community colleges.

At the summit, Jane Oates, assistant secretary of the Employment Training Administration at the U.S. Department of Labor, described the department's interest in community colleges. The Department of Labor has a particular concern with students who did not make it through high school and with adults who need to go back to school because they have lost employment. Many people who walk onto a community college campus would not feel comfortable going to a four-year public or private college, said Oates. They may be people who have worked in a single industry like the auto industry or general manufacturing for many years, no longer have a job, and are eligible for training. "The people I work with every day need a job as soon as possible, and they need a job where they can continue their education on a career pathway," said Oates.

Before 2009, the Department of Labor did not insist that the training it funded lead to an industry-recognized credential or a pathway to a degree. The department also limited community colleges' use of federal funds for equipment, though equipment is essential in many educational fields. Finally, the department did not require rigorous evaluations.

Beginning in 2009, the department has been addressing those weaknesses, said Oates. Under the Trade Adjustment Assistance Community College and Ca-

---

introduction to the discussions summarized in Chapter 7. Chapter 7 of this report gathers comments made by summit participants on a variety of issues affecting the future of community colleges.

BUILDING ON THE INTEREST IN COMMUNITY COLLEGES

A meeting planned for 60 to 75 participants quickly grew to more than 100. In addition, more than 150 people registered to watch a webcast reer Training Grant Program,[a] it has begun awarding what will be a total of $4 billion over two years to help prepare students for successful careers in growing and emerging industries. It also began rigorous evaluation of grants and formula funds. In the first round of capacity-building grants, consortia of institutions received about 60 percent of the $500 million distributed in 2011. These consortia mostly formed around needs of particular sectors such as advanced manufacturing, healthcare, and engineering. They not only developed new curricula based on the needs of employers but looked at new methods of delivering educational content, such as online learning. Community colleges have not been in the forefront of online educational innovations, said Oates, but they are the point of entry for many people who could benefit from such learning.

The Department of Labor has been supporting the development of other electronic tools. A web-based tool called My Skills My Future,[b] which received more than 2.5 million hits over 14 months, allows people to see jobs that are currently available. Similarly, a tool called My Next Move[c] allows dislocated workers to search by zip code for jobs that they have held before in their local area or to match jobs with additional skills that they have. My Next Move for Veterans[d] allows ex-military personnel to crosswalk their military job codes with civilian job titles. All of these sites draw heavily on the Labor Department's partnerships with community colleges to find matches with education and with jobs.

Finally, the Workforce Innovation Fund,[e] in partnership with the U.S. Department of Education, is highlighting the innovations and partnerships between the workforce and community colleges. "The time is right for us to talk about the rigor and the wonder and the innovation that are going on in community colleges," said Oates. "We have for too long seen them as a stepchild, and they can do amazing things."

---

[a]For additional information, see http://www.doleta.gov/taaccct/.
[b]Available at http://www.myskillsmyfuture.org/.
[c]Available at http://www.mynextmove.org/.
[d]Available at http://www.mynextmove.org/vets/.
[e]Available at http://www.doleta.gov/workforce_innovation/.

---

of the meeting.[4] Attendees included representatives of the White House Office of Science and Technology Policy, NSF, the U.S. Department of Education, the U.S. Department of Labor, state education agencies, private foundations, businesses, higher education organizations, faculty and administrators from two-year and four-year colleges and universities, several students from local community colleges, and K-12 teachers. It was a diverse, energetic, and enthusiastic group.

---

[4]Video archives of the presentations are available through a link on the summit's website (see http://nas-sites.org/communitycollegessummit/).

Responses to a Pre-Summit Survey:
Challenges in STEM Education and Careers

As part of the registration process to attend the National Academies' Summit on Community Colleges in an Evolving STEM Education Landscape, invited participants were asked two questions:

1. What is the greatest challenge or issue you are facing in your work on two-year or four-year STEM education and careers?
2. What is the one big idea or insight you have about increasing the potential of community colleges in STEM education and careers that you will bring to the summit?

Participants' big ideas and insights are summarized in Chapter 7. The challenges or issues they listed most frequently are the following:

1. **Overcoming students' inadequate academic preparation for STEM study**. Students interested in pursuing a STEM career often begin their two-year or four-year study with too little preparation in mathematics, reasoning, and critical thinking to succeed. Extra coursework is required for remediation, which lengthens the time for earning a degree and slows academic progress. As a result, students become discouraged from pursuing a STEM career and either change the focus of their studies or fail to complete their degree.
2. **Recruiting and retaining students in STEM education**. Outreach, inspiring students to pursue a STEM career, and then helping them overcome barriers along the way, all pose significant challenges. This is especially true for women and minorities.

The National Academies have served as an interactive forum for people who should be talking with each other but do not often have opportunities to do so. In this regard, the summit was a great success. Participants listened to each other, reflected on new ideas and insights, learned about pressing issues, and thought about how those issues affect their own domains. As George Boggs, former president of the American Association of Community Colleges, a member of the NRC's Board on Science Education, and the chair of the summit organizing committee, said in his concluding remarks, "This summit is the start of something that will be beneficial for all of our institutions, for our students, and for the country."

3. **Creating and sustaining effective partnerships between two-year and four-year institutions.** Although individual two-year and four-year institutions in some regions and states have forged effective STEM education partnerships, these partnerships could be much more widely implemented and reflected in state and local policies. In particular, four-year institutions need to have a greater appreciation for the kinds of modern approaches and subject matter in STEM education that are offered at two-year colleges. Partnerships are needed to address such issues as curriculum and other program alignment issues, getting staff and faculty at both institutions on board with student needs and program requirements, and providing course and program articulation policies and practices between two-year and four-year institutions. Partnerships are also essential in developing pathways from a technical degree into a full baccalaureate, especially if some time has passed since the student has completed an associate's degree. In general, transfer and articulation policies and practices are frequently mentioned barriers to retention in STEM education.

4. **Finding the resources to support and sustain STEM education program improvement.** There is a universal lack of time and dependable, sustainable resources to support the necessary STEM education collaborations and program improvement initiatives. Furthermore, the weak economy has had a major impact on those efforts. Both two-year and four-year institutions struggle with the high cost of laboratory facilities. In addition, community colleges have difficulty in obtaining and managing external funding.

5. **Aligning STEM education with workforce demands and practices.** The academic and corporate agendas for STEM education that enable students to advance from two-year to four-year degrees in these fields and the need to offer programs that propel students toward specific careers in STEM are not always well aligned. In general, the dual role of community colleges to educate students for careers and matriculation in four-year programs is a continuing tension for community colleges.

The White House Summit

The administration of President Barack Obama has devoted considerable attention to community colleges. As part of his call to increase college enrollments and completion among young people, the President has asked community colleges to increase the number of graduates and program completers by five million students over a 10-year period, representing a 50 percent increase over current numbers.

In October 2010, the Obama Administration held a summit on community colleges at the White House organized by Jill Biden, the wife of Vice President Joseph Biden and an adjunct professor of English at Northern Virginia Community College. At that summit, President Obama called community colleges the "unsung heroes" of American education and emphasized the critical role they play in sustaining the nation's competitiveness. He pointed out that in the coming years jobs requiring at least an associate's degree are projected to increase twice as fast as those requiring no college experience. "We will not fill those jobs—or keep those jobs on our shores—without community colleges," the President said (White House, 2011).

2

Expanding Minority Participation in Undergraduate STEM Education

---

Important Points Made by the Speaker

- Community colleges need to be integrally involved in a comprehensive, coordinated, and sustained effort to increase the participation of under-represented minorities in STEM education and careers.
- Increasing the completion rate of underrepresented minorities in STEM majors requires a combination of strong academic, social, and financial support.
- To increase the number of U.S. students who earn degrees in STEM fields, all institutions of higher education must work to create a culture in which it is "cool to be smart."

The demographics of the U.S. population are undergoing a dramatic shift, observed Freeman Hrabowski, president of the University of Maryland, Baltimore County (UMBC), in his keynote speech at the summit. Minority groups underrepresented in STEM fields soon will make up the majority of school-age children in the United States (Frey, 2012). To maintain the strength and vitality of science and technology in the United States, many more of these minority children must not only decide to become scientists and engineers but succeed in educational pathways that allow them to do so. Given the overrepresentation of minority students in community colleges, community colleges will be critical in achieving this goal.

Drawing from the recent report of a committee that he chaired

(National Academy of Sciences, National Academy of Engineering, and Institute of Medicine, 2011), Hrabowski pointed out that the proportion of underrepresented minorities in the natural sciences and engineering was less than a third of their share of the overall population in 2006 (National Science Foundation, 2011). In other words, the proportion of underrepresented minorities in science and engineering would need to triple to match their representation in the overall U.S. population.

This underrepresentation of minorities in the science and engineering workforce stems from the underproduction of minorities in science and engineering at every level of the pathways from elementary school to higher education and the workplace. Though underrepresented minorities now account for almost 40 percent of K-12 students in the United States, they earn only 27 percent of the associate's degrees from community colleges, only 17 percent of the bachelor's degrees in the natural sciences and engineering, and only 6.6 percent of the doctorates in those fields.

President Obama has called on the United States to increase its postsecondary completion rate from 39 percent to 58–60 percent by the year 2020.[1] The challenge in doing so is greatest for minorities who are underrepresented in science and engineering. According to 2006 data, of Americans aged 25 to 34, only about one quarter of African Americans, Native Americans, and Pacific Islanders had earned at least an associate's degree, and fewer than one in five Hispanics had reached this educational level.

In 2000, the United States ranked 20th in the world in the percentage of 24-year-olds who had earned a first college degree in the natural sciences and engineering, Hrabowski noted. The report *Rising Above the Gathering Storm* (National Academy of Sciences, National Academy of Engineering, and Institute of Medicine, 2007) called on the United States to raise the percentage of 24-year-olds with a first degree in the natural sciences and engineering from 6 percent to 10 percent. This would require a tripling, quadrupling, or quintupling of the percentages for underrepresented minorities, which are 2.7 percent for African Americans, 3.3 percent for Native Americans, and 2.2 percent for Latinos.

INTENTIONS AND COMPLETIONS

Since the 1980s, underrepresented minorities have aspired to major in science and engineering at about the same proportions as their white and Asian American peers, Hrabowski observed. Yet they complete STEM degrees in lower proportions than whites and Asian Americans. Five

---

[1]For additional information, see http://www.whitehouse.gov/sites/default/files/completion_state_by_state.pdf.

years after matriculating, only about 20 percent of underrepresented minorities who intended to earn a STEM degree have done so. Surprisingly, only about one-third of whites and slightly more than 40 percent of Asian Americans earn STEM degrees within five years.

Hrabowski ascribed part of this attrition to the culture of science and engineering in college. A large part of the problem is the "weed-out" mentality still held by many college faculty in these subjects, he said. When students have difficulties with their initial classes, they are more likely to be encouraged to transfer to another major than to receive help in overcoming those difficulties. Hrabowski recounted talking to the directors of the institutes at the National Institutes of Health (NIH) and saying that he had many friends who started in science or engineering and became great lawyers. "Everybody laughed, but afterwards the General Counsel of NIH came to me and said, 'You just told my story. I went to one of the Ivies. I started off in science. I had the best of test scores, the best of grades. I got wiped out in the first year and I changed to pre-law.' It happens all the time." Not only do such experiences lead to fewer students of all races majoring in science or engineering, but also they affect attitudes in general toward the subjects. He said, "You have to ask, how could Americans really love science or math . . . if they started off and [ended] getting wiped out? There is a negativity. When I ask audiences, 'How many of you love to read?' everybody raises their hand. Then I ask, 'How many of you love math?' and people start to laugh."

POLICY INITIATIVES

The problem is urgent, Hrabowski said. A national effort to address underrepresented minority participation and success in STEM fields needs to be initiated and sustained. This effort must focus on all segments of the pathways, all stakeholders, and the potential of all programs, whether targeted at underrepresented minorities or at all students. Students who have had less exposure to STEM and to postsecondary education than others require more intensive efforts at each level to provide adequate preparation, financial support, mentoring, social integration, and professional development. Evaluations of STEM programs, along with increased research on the many dimensions of underrepresented minorities' experiences, are needed to ensure that programs are well informed, well designed, and successful.

The NRC committee that Hrabowski chaired made recommendations in *Expanding Underrepresented Minority Participation: America's Science and Technology Talent at the Crossroads* (National Academy of Sciences, National Academy of Engineering, and Institute of Medicine, 2011) at the preschool through grade 12 level in the areas of early readiness, mathematics and science instruction, and teacher preparation and retention. At the summit, however, Hrabowski focused his comments on the postsecondary level. Underrepresented students need improved access to postsecondary education and technical training. They need more awareness of and motivation to pursue STEM education and careers. They also need adequate financial support. "It is impossible for a student to do well in biochemistry while working 25 hours on the outside," said Hrabowski. "When you are doing all of that work on the outside, it is almost impossible to succeed in science."

Colleges and universities need to institute reforms to increase the inclusiveness and success of underrepresented students in STEM fields. Colleges have a tendency to say that the problem is at the K-12 level, but Hrabowski disagreed. He said K-12 education does need to be strengthened, but more students are better prepared than many faculty and administrators at colleges and universities think. According to Sylvia Hurtado, who directs the Higher Education Research Institute at the University of California, Los Angeles, and was on the NRC committee Hrabowski chaired, the larger the number of Advanced Placement credits a student has taken, the higher the SAT, and the more selective the university, the greater the probability the student will leave science as an undergraduate, noting "It is not just a matter of preparation." When college presidents point out to Hrabowski that most of the underrepresented students interested in science and engineering leave these majors, he responds that the majority of white and Asian American students do, too.

The NRC (2011) committee recommended increasing the completion rate of underrepresented students by providing strong academic, social, and financial support. This support should come from programs that simultaneously integrate academic, social, and professional development. Programs also are needed that facilitate the transition from undergraduate to graduate education and provide support for graduate students.

THE CHALLENGE FOR TWO-YEAR INSTITUTIONS

Hrabowski cited several challenges that are particularly acute for community colleges. Inadequate levels of mathematical preparation are a problem for almost all colleges and universities, but it is an especially difficult problem at community colleges. (This issue is the subject of Chapter 5.) Community colleges also need to balance the tasks of preparing students for further study at four-year colleges and graduate schools along with preparation resulting in two-year degrees and other certificates for the technical workforce. To facilitate and increase the transfer of underrepresented students in STEM to four-year institutions, increased emphasis and support are needed for articulation agreements, summer bridge programs, mentoring, academic and career counseling, peer support, tutoring, social integration activities, study groups, undergraduate research, and tracking of student progress. (Transfer issues are discussed in Chapter 6.)

Several federal programs facilitate the transfer of underrepresented minorities from community colleges to four-year institutions, Hrabowski noted, such as the Bridges to the Baccalaureate[2] Program and the Community College Summer Enrichment Program at NIH.[3] Community colleges also have mounted such promising initiatives as Miami Dade College's Windows of Opportunity Program,[4] which helps academically promising, low-income students in obtaining associate's degrees in the arts or in STEM disciplines. Strategies that promote transfer include grants that allow community college students to work less outside of their academic programs and complete their associate's degrees in three years and then successfully transfer to complete their four-year degrees.

The Gates Foundation is supporting research in the state of Maryland to track students who have been majoring in science and pre-engineering areas and look at what happens to them when they come to four-year institutions. The underlying objective is to get faculty at different institutions to talk honestly and openly about how work at one level is related to work at the next level. Even if courses have the same name, they may not be at the same level, so students who transfer to a four-year college are not as prepared as they need to be, Hrabowski stated. The initiative also gives faculty at the four-year institutions a better idea of the challenges that community colleges are facing.[5]

Finally, Hrabowski emphasized the potential for internships to motivate students and prepare them for careers. "When students have internships . . . and see the connection between their academic work and what is going on in a company, they get even more excited," he said. Internships make students more serious about their work. The needs of industry can be infused into the curriculum, especially when people from business are involved in developing or teaching the courses. Students learn how to work in teams, express themselves clearly, and gain other 21st century

---

[2] Additional information is available at http://www.nigms.nih.gov/Research/Mechanisms/BridgesBaccalaureate.htm.

[3] Additional information is available at https://www.training.nih.gov/ccsep_home_page.

[4] Additional information is available at http://www.toolsforsuccess.org/#SlideFrame_1.

[5] A website for this program was not available at the time that this report was prepared. Additional information about this program is currently available as a downloadable PowerPoint presentation titled "A Shared Responsibility: Creating a t-STEM Friendly Multi-Campus Community (T-STEM Cross-Campus Collaboration Team, UMBC)" at http://transferinstitute.unt.edu/content/10th-annual-conference-national-institute-study-transfer-students.

skills[6] that they can use in the workplace (e.g., National Research Council, 2010). In the area of cybersecurity, to cite just one example, Hrabowski noted that the majority of students who have internships go to work full time for the same companies once they graduate. "It is amazing how much more students will do when they get connected to the company early."

PROGRAMS AT UMBC

Hrabowski mentioned several programs at UMBC, which is nationally recognized for its Meyerhoff Scholars Program,[7] that involve community colleges. For example, UMBC has a Chemistry Discovery Center[8] that is working with community colleges with an emphasis on group work, use of technology, collaboration, and professional development for faculty. The key, said Hrabowski, is to involve people in an activity that they see as exciting. "That is the thing about my campus," he said. "We are seeing amazing results for students of all races where the emphasis is on student engagement and on empowering students."

UMBC has memoranda of understanding with four community colleges in Maryland, which enable both students *and* faculty to move back and forth among the institutions, and similar arrangements could be made at many four-year and two-year colleges. The Gates grant is also producing more communication and movement among institutions.

UMBC has contacts with many companies in such areas as biotechnology, computer security, defense, and environmental protection that are very interested in hiring students not just at the PhD level but at the four-year and two-year levels as well. High school counselors, families, and students all need to know about these options, which provide what Hrabowski called "great jobs, good-paying jobs," and about how to take advantage of them.

UMBC enrolls students from 150 countries, and those international students are hungry for knowledge. Having them on campus makes the U.S. students focus and push harder, according to Hrabowski. Through-

---

[6]NRC (2010, p. 3) lists as "21st century skills" adaptability, complex communication/social skills, non-routine problem solving, self-management and self-development, and systems thinking. NRC (2011, p. 1) further refines this list to include "… being able to solve complex problems, to think critically about tasks, to effectively, communicate with people from a variety of different cultures and using a variety of different techniques, to work in collaboration with others, to adapt to rapidly changing environments and conditions for performing tasks, to effectively manage one's work, and to acquire new skills and information on one's own."

[7]Additional information is available at http://www.umbc.edu/meyerhoff/.

[8]Additional information is available at http://www.umbc.edu/chem/facilities/discovery.html.

out the 20th century, many of the best figures in science and engineering had parents who came from other countries. Today, the majority of black students on Ivy League campuses have parents from another country. "It has everything to do with the hunger for the knowledge," Hrabowski said. "You have to work really hard."

DISCUSSION

During the discussion period, Hrabowski remarked on some of the factors behind the success of the Meyerhoff Program, which is the national leader in producing African American graduates at a predominantly white school who go on to complete their PhD in science and engineering, with 12 to 15 of these graduates typically earning a STEM PhD each year. The program has created a culture where it is "cool to be smart," Hrabowski said. Multiple academic and social connections link students to each other, to faculty members, and to community members. Students are engaged in projects rather than just sitting in lectures, which has required that courses be redesigned. Students are connected with companies through classroom projects and internships to show them what it takes to get a good job. Graduates who get these jobs in turn come back and talk with students. Recent graduates and current students know better than anyone else how to get more students involved.

Hrabowski has written books on raising smart black children (Hrabowski et al., 2002; Hrabowski, Maton, and Greif, 1998), and he emphasized what parents in successful families do: work with their children to develop their reading, thinking, and studying skills. Succeeding in mathematics or science is not always fun, and he emphasized, "Hard work is hard work. We have to get students engaged in the work."

Hrabowski also noted during the discussion session that completion of degree programs is a problem at community colleges. "Anybody from a community college knows what I am saying. Most of the kids who start off saying they want something at a two-year college do not finish the program," he said. NSF should support an effort to assess what percentage of community college students who start in STEM programs finish them, Hrabowski urged.

Faculty and staff involvement is critical. Many faculty are not familiar with the data regarding contributors to student performance. In making decisions about education, they tend to rely on anecdotal or impressionistic information, according to Hrabowski. Effective interventions require that actions be based on data.

Summit participant Rebecca Hartzler, now at the Carnegie Foundation for the Advancement of Teaching, pointed out that the numbers of women and underrepresented minorities in some fields might need to increase by an order of magnitude for representation to be proportional. She was the first tenured woman in physics in community colleges in Washington State, and there may now be two more. "We don't want three physics faculty in 34 community colleges in Washington State," she said. "We want 30 women teaching physics in Washington State. We [need] much larger ambitions."

3

The Loss of Students from STEM Majors

Important Points Made by the Speaker

- More than half of students who intend to major in a science or engineering field switch to a different major in college, and this percentage is even higher for community college students.
- Students who take fewer STEM classes their first semester are more likely to switch from STEM majors.
- The culture of STEM education and potential earnings in the workplace appear to be significant factors in students' decisions to remain in or abandon STEM majors.

Students who intend to work as scientists, technicians, engineers, or mathematicians typically choose to major in a STEM field in college. Study of the choices students make regarding majors both before and during college therefore can reveal important information about the future U.S. workforce.

Eric Bettinger, associate professor for education and economics at Stanford University, has used data from Ohio to analyze students' choices of majors and how those decisions change over the course of a two-year and four-year education (Bettinger, 2010). His data are from the 1998–1999 cohort of incoming students, which allowed him to follow their choices in subsequent years, and the data focus solely on students who took the ACT exam, which is the exam taken by most college-bound students in Ohio. Students who take that exam indicate the major they would like to pursue, and the exam results allow high-ability students to be identified and analyzed separately.

In a total sample of 18,000 students, 8.0 percent and 11.7 percent indicated an interest in the biological or physical sciences and engineering, respectively. Bettinger found, in an analysis conducted for the summit, that these numbers were somewhat lower for the students who attended two-year colleges—5.5 percent and 9.4 percent. In contrast, the percentages were higher for students with high ACT scores (above 24)—11.7 percent and 18.0 percent. The students at two-year institutions had somewhat lower average ACT scores than the average for all students, but in broad terms their aspirations and characteristics were similar, Bettinger said.

STUDENTS WHO LEAVE AND ENTER STEM MAJORS

A "depressing" number of students abandon STEM majors, Bettinger observed. Among all students who declared an intention to pursue a STEM major, only 43 percent were still in a STEM field at the time of their last enrollment, with the rest moving to other majors by the time of their last enrollment.

The numbers were far worse for two-year students. Only 14 percent of the students at two-year colleges who intended to major in a STEM field when they took the ACT exam were still in a STEM field at the time of their last enrollment. "This defection rate is extremely high," said Bettinger.

Almost one-half of all students who leave STEM majors switch to business majors (48.7%). Other popular majors for students who switch are the social sciences (21.2%) and education (11.1%). Among two-year switchers, about 30 percent switch to business majors, and slightly less than one-quarter each go to social science and education majors.

Meanwhile, very few students who did not intend to major in a STEM subject converted to a STEM major. Only 5.5 percent of STEM majors for students at all institutions, and only 3.4 percent for two-year students, were converts to STEM from a non-STEM major.

WHY DO STUDENTS LEAVE STEM MAJORS?

Bettinger listed five possible reasons for the relative lack of U.S. students pursuing STEM majors in two-year and four-year institutions:

1. At the end of secondary school, few are prepared to enter STEM fields.
2. Few express initial interest in entering STEM fields.

3. Once students are off the STEM pathway, they cannot get back on it.
4. The culture of STEM fields is off-putting once higher education is reached.
5. The returns are insufficiently high to justify greater adherence to STEM fields.

He noted that his data are best suited to explore the last three of these explanations. Students started switching away from STEM majors in their very first semester, and the students most likely to leave STEM majors were the ones who took fewer STEM courses their first semester rather than more courses. Students who took more than 40 percent of their courses in STEM their first semester were much less likely to leave the major than students who took less than 40 percent of their courses in STEM fields. This observation holds for students in four-year colleges, students in two-year colleges, and high-ability students.

The relatively small number of students who converted to STEM majors also took relatively few STEM courses their first semester. This piece of evidence is "suggestive," said Bettinger, that there might be some way of getting more non-STEM majors interested in those subjects—for example, by examining more closely the structure and conduct of introductory courses in STEM. However, STEM majors have extensive course requirements, and many courses typically must be taken in a particular order (that is, they have extensive prerequisites compared with other disciplines), which can make it difficult to switch into these majors.

The students who left STEM were just as likely to pass their initial STEM courses, so the difficulty of the courses did not seem to be the deciding factor. But the course demands of STEM majors are high and require commitment—even though, as Bettinger observed, some of the majors to which students switch, such as education, also have extensive course requirements, even if they are not as sequential as those for STEM majors.

THE CULTURE OF STEM FIELDS

Bettinger's data also show that women were significantly less likely to stay in STEM fields, even among the top students, which suggests that the culture of STEM might have been a factor in their decisions. However, since the female students took STEM courses in high school and still expressed an interest in majoring in those subjects, the cultural problems would need to start or intensify in college for this explanation to hold.

According to Bettinger's research, black students in four-year colleges were less likely to defect from STEM majors than other students, especially among the top black students. However, that was not true at two-year colleges, where there were no statistical differences between black students and other students. Bettinger did not analyze the differences in these indicators between men and women or between domestic and international students, though both of these factors could influence the results.

EARNINGS FROM MAJORS

One factor in students' decisions about majors is the amount of money they potentially could earn after graduation. About three-quarters of college students respond in surveys that an important objective of a college education is to be "well off financially" (Pryor et al., 2011), and colleges have an increased focus on vocational offerings, particularly at two-year colleges.

The data suggest that students who switch to a non-STEM major could have been making a calculated decision about where the financial return to a major might be higher than with a STEM major, Bettinger said. For example, women's earnings in business and in other fields were higher than they were in STEM fields at the time these data were gathered, though men's earnings in STEM fields, business, and the social sciences were roughly the same.

High average earnings indicate similarly high levels of demand for workers, and superstar earnings indicate a demand for a large pool of professionals to produce a small number of superstars. For example, computer science, which is a field with obvious earnings growth and superstar earnings, was experiencing a substantial growth in majors at the time the data were gathered, Bettinger noted.

DISCUSSION

During one of the discussion periods at the summit, Catherine Didion from the National Academy of Engineering pointed out that underrepresented students and women are interested in giving back to their communities but often do not see STEM fields as occupations that enable them to do so. Additional investigations could indicate why so many of these students switch into non-STEM fields.

Martha Kanter, under secretary at the U.S. Department of Education, emphasized the importance of mentors and advisers in keeping students on track. Many students take courses they do not need, or they have unclear pathways. Students need sophisticated and knowledgeable advice. "Students get lost in the system," said Kanter. "We have to use technology and people to keep them in the system and keep them highly motivated to succeed."

4

Outreach, Recruitment, and Mentoring

Important Points Made by the Speaker

- Many students lack the information they need to succeed in community college, especially if they are interested in pursuing a STEM degree.
- Students need to be exposed to STEM occupations and learn what they need to do to qualify for those occupations.
- Hands-on programs designed to recruit students into STEM fields need to be paired with academic preparation.
- Grants to colleges that require undergraduate mentoring plans could give more students the information they need to persist in STEM education.

AN ECOLOGICAL MODEL

In summarizing the main messages of the background paper she prepared for the summit (see Appendix B), Becky Packard of Mount Holyoke College used an "ecological model" that examines the many environmental factors and the relationships among those factors that affect students' choices. Students like Josie, described at the beginning of Chapter 1 of this report, are influenced by the home, the school, the workplace, and other contexts. Their access to resources, transportation, financial aid, and child care informs their choices. Many students, including first-generation and low-income students, do not gain the knowledge they need to navigate the college application and enrollment process successfully. Financial considerations can be and often are a significant barrier to college entrance and persistence. And students often lack information on transfer requirements and what they are likely to experience if they do transfer.

When students gain mentoring from multiple contexts, they are more likely not only to persist in college but to do so in a STEM major, said Packard. "We have a pretty good idea of what we should be investing in," she said.

OUTREACH

Packard highlighted several of the recommendations from her background paper as particularly important to the summit.

First, more students and families need to understand the difference between a technical degree from a career institute and the community college transfer pathway to a four-year STEM degree. They also need to know much more about the STEM careers that are available. Students should be exposed to STEM occupations and learn what they need to do to qualify for those occupations.

Excellent models already exist, Packard observed. Statewide and nationwide, programs have developed coordinated approaches to outreach so that messages from high schools, community colleges, and four-year institutions are reinforcing. The Advanced Technology Education centers funded by the National Science Foundation (NSF)[1] and other programs to broaden participation also have outreach programs in place.

RECRUITMENT

In the area of recruitment, any hands-on program designed to attract students into STEM needs to be paired with academic preparation. Specifically, Packard suggested an expansion in STEM-specific dual-enrollment programs with community colleges or universities while students are in high school in addition to more common outreach and recruitment strategies such as summer enrichment programs. "Summer enrichment programs can enhance interest or get students to take a first course," said Packard, "but greater academic preparation makes it realistic for them to continue going forward."

Both honors students and struggling students can benefit from dual-enrollment courses because taking college classes during high school can motivate students to continue their education. In addition, high school students should be able to count their college classes for a

---

[1]Additional information is available at http://www.nsf.gov/pubs/2011/nsf11692/nsf11692.htm and http://atecenters.org.

high school requirement, which would allow students with lower grade point averages to take advantage of dual-enrollment classes.

Packard also called attention to the deterrent posed by developmental mathematics, which is discussed in the next chapter. Experiential programs and dual-enrollment classes that target academic mathematics requirements could recruit students into STEM fields and give them a realistic chance of persisting.

MENTORING

Research on mentoring is robust, sophisticated, and rigorous, Packard noted. Most of the newer studies are comparative, longitudinal, or control for self-selection issues. However, more research is needed on how to create more effective mentoring programs and bring effective mentoring programs to scale (references are provided in Appendix B).

Packard also recommended that grants require undergraduate mentoring plans. NSF has such a requirement for postdoctoral researchers, and there is no reason why this provision could not be extended to undergraduate students, she said.

Finally, she said, informal mentoring and advising need to be infused by faculty into all courses. Mentoring cannot be done through supplemental programs alone.

PRODUCTIVE INVESTMENTS

The American Institutes for Research has estimated that more than $4 billion in grants and state allocations are lost when new, full-time community college students do not return for a second year of study (Schneider, 2011). According to the report in which this estimate appeared, said Packard, "The only thing more expensive than fixing retention in community college is not fixing it."

Packard was part of the first generation in her family to complete a four-year degree. A summer research experience motivated her to get a PhD, which led to research support from NSF and a Presidential Early Career Award for Scientists and Engineers. "Every single day I am grateful for the professor mentor who got me into a carpool and was flexible enough so that I could work my other two jobs," she said. Yet the challenges that she faced pale in comparison to those faced by many students who are trying to navigate the community college transfer pathway. "Students are not just data sources to me," she said. "I am deeply troubled by the struggles that students face when trying to navigate these pathways to four-year STEM degrees."

Collective Observations from a Breakout Group on Outreach

From the breakout discussion on outreach issues, participants reported to all Summit attendees on three main messages:

First, mentoring and role modeling could be expanded to encompass colleagues and partners in business and industry. For example, professionals, especially those with backgrounds and experiences similar to those of their student audiences, might visit high schools and help inspire students to pursue STEM careers. Students who are more knowledgeable about what they can do with STEM skills can then work with advisors to develop plans for developing those skills.

Second, both students and faculty would benefit from more accurate and honest information about educational pathways. Such information could be disseminated more strategically and systematically. The same information could then be disseminated to students at both four-year institutions and community colleges so that they do not receive conflicting advice.

Third, more research is needed to validate existing programs, and the results of this research need to be communicated, especially to business and industry. Community colleges remain a well-kept secret in many communities. They may communicate well among themselves, but they do not always communicate well with the rest of the world. For example, it could be helpful for congressional policy makers to understand what the lives of community college students are like so that legislation and appropriations might be better tailored to their lives and needs.

DISCUSSION

In response to a question from George Boggs about orientation for transfer students, Packard said that some institutions have done a good job, though they are still the exceptions. Furthermore, she emphasized that transfer students should receive not just orientation but discipline-specific orientation. Students need more than just a general acclimation to a college. They need to know what they missed in their education compared with better-prepared students so that they are not lost when pre-existing knowledge is taken for granted. University administrators and faculty members need to put themselves in the shoes of transfer students and ask how they would fare. "It is a huge eye opener," Packard said. "It is not just that the students are unintelligent."

Linnea Fletcher from Austin Community College, who was a member of the organizing committee, said that as soon as she was away from community colleges for several years as a rotator at NSF, she started to forget what it is like to be a student at a community college. "As soon as you lose that perspective, you no longer can connect with those students and understand what their lives are like," she said. For example, Fletcher pointed out that scholarships requiring students to take a full load and not work are unrealistic for many students at community colleges, such as students with families. All levels of education must consider what it takes to make education work for their students, Fletcher said. In addition, Judy Miner from Foothill College, who was also a member of the organizing committee, reminded the participants of the importance of family engagement, community, and culture in attracting and retaining students of color in STEM.

Another member of the organizing committee, Thomas Bailey from Teachers College, Columbia University, observed that one potent recruiting tool for STEM transfer students could be introductory science courses at community colleges. Developmental education courses or student success courses are other ways to help students find direction. Students often are not aware of the requirements for transferring, even when they have declared a major leading to eventual transfer. "What is our responsibility to make sure that those students are progressing in a systematic and coherent way toward those goals?" asked Bailey. More advisers are needed, of course, but community colleges will need to combine technology with advisers given their financial constraints.

5

The Two-Year Curriculum in Mathematics

Important Points Made by the Speaker

- Many more students are taking mathematics at community colleges than has been the case in the past, but the majority of students enroll at the precollege, noncredit level.
- Reform of the mathematics curriculum needs to encompass the entire educational system.
- Much more research is needed on teaching and learning in two-year college mathematics and on the characteristics, experiences, and aspirations of students.
- Practitioners need to be engaged in research on mathematics education to facilitate adoption and scale-up.

Mathematics is seen by many as the backbone of the STEM pipeline, said Debra Bragg, professor of higher education at the University of Illinois, and author of one of the three papers commissioned for the summit. The complete paper is available in Appendix C. Yet very few students in community colleges ever progress beyond arithmetic or algebra. Though reforming mathematics education in the United States is an "enormous" job, said Bragg, changes at the community college level can help set the process in motion.

RISING ENROLLMENTS

The normative mathematics sequence in U.S. education progresses from arithmetic to algebra to geometry to trigonometry to calculus. Over the past three decades, many more students have at least embarked upon this progression in two-year institutions—from about one million students enrolled in two-year mathematics and statistics programs in the early 1980s to more than two million today, according to data provided to Bragg from the Conference Board of Mathematical Sciences. Furthermore, about 47 percent of mathematics enrollments in higher education are at the two-year level. "That is a lot of enrollments and clearly a very important part of the pipeline," said Bragg.

However, 57 percent of the students enrolled in two-year college mathematics are enrolled at the pre-college, noncredit level. The course with the largest enrollment is elementary algebra, which is usually one to two levels below college-level algebra. Over the past five years, the greatest growth in enrollments has been in arithmetic and pre-college algebra. "We are seeing growth at the lower end, not where we were hoping to see it," she said.

The preponderance of enrollments in college-level mathematics is in college algebra, and most students do not move beyond that level. Only about 7 percent of enrollments are in calculus, and only about 7 percent are in statistics, with most students never moving beyond the introductory courses in these subjects. Other significant enrollments are pre-calculus (18%) and other mathematics classes (11%) such as linear algebra, mathematics for elementary teachers, or non-calculus mathematics for technical careers.

INSTRUCTIONAL APPROACHES

The Conference Board of Mathematical Sciences also has conducted a survey about instructional approaches in two-year mathematics courses.[1] Relatively few two-year courses offer special mathematics programs that provide support for minorities or women (11% and 6%, respectively). About 14 percent offer undergraduate research opportunities, and 20 percent offer honors sections to mathematics students.

In contrast to these sparse offerings, 90 percent of two-year college mathematics programs require diagnostic or placement testing. An increasing number of researchers are raising questions about the use of

---

[1]The survey is available at http://www.ams.org/profession/data/cbms-survey/cbms2005.

these tests, said Bragg, and about alternative educational approaches that could reduce the number of students needing developmental mathematics.

MATHEMATICS REFORM

The American Mathematical Association of Two-Year Colleges (AMATYC) has made an extended commitment to reform in mathematics education. The AMATYC Crossroads in Mathematics Program[2] led to follow-up programs called Beyond Crossroads[3] and College Renewal Across the First Two Years, under the aegis of the Mathematical Association of America,[4] which have tackled the implementation challenges inherent in reform. In addition, work by Lynn Steen, Uri Treisman, and others have contributed to careful thinking about what and how mathematics is taught, Bragg said (references are in Appendix C).

SPEAKER AND PARTICIPANT SUGGESTIONS FOR FUTURE ACTION

Bragg made four suggestions for future action on the basis of her observations.

First, reform of the mathematics curriculum needs to encompass the entire educational system. Without a strategic, collaborative endeavor, it will be difficult for two-year colleges, caught as they are between K-12 education and universities, to implement and sustain reform, except in isolated ways. Today, reform at different levels is largely separate, Bragg said; it needs to be combined and integrated.

Second, much more research is needed on teaching and learning in two-year college mathematics, especially in college-level mathematics. Numerous pedagogical strategies are emerging that have promise to change the way two-year college mathematics is taught, said Bragg, but today lecture-led, teacher-centered instruction predominates.

Third, the characteristics, experiences, and aspirations of students who enroll in two-year college mathematics need to be investigated in greater depth. More research is needed to understand how students develop the "habits of the mathematical mind" that are required to be successful in all STEM fields.

Finally, practitioners need to be engaged in research on mathematics education to facilitate adoption and scale-up. Two-year faculty would

---

[2] Additional information is available at http://www.amatyc.org/Crossroads/CrsrdsXS.pdf.

[3] Additional information is available at http://beyondcrossroads.amatyc.org/.

[4] Additional information is available at http://www.maa.org/cupm/crafty/.

appreciate and benefit from opportunities to engage in research that encourages them to try out new pedagogical strategies in the classroom and determine how they affect student learning. "The math faculty will be hungry and excited to be part of this kind of research, because they live this issue every day," Bragg said.

Report of Collective Observations from a Breakout Group on Mathematics

Participants in the breakout session on mathematics education at the summit reported three main observations from their discussions during a plenary session of all Summit attendees:

First, additional research about mathematics education at the community college level could lead to more informed policies and decision making.

Second, successful evidence-based instructional systems for mathematics need to be identified. Research on instruction indicates that effective systems encompass curriculum, pedagogy, faculty development, and student support mechanisms.

Third, excellent evidence-based instructional systems, which combine the research and identification of pockets of excellence, exist today. However, there are too few documented cases where they are being strategically replicated and expanded.

DISCUSSION

During the discussion sessions at the summit, participants made a number of comments related to mathematics at the community college level.

Pamela Brown from the National Science Foundation, on leave from the New York City College of Technology, a branch of the City University of New York, directed attention to the 60 percent of the institution's 16,000 incoming students who need to take developmental mathematics. "I would not describe it as a gatekeeper," she said. "I would have to say it is more like a firing squad, because only about 20 percent of the students pass the lowest levels of developmental math, and a great percentage of those students withdraw unofficially. They just give up and stop coming to classes." Part-time faculty who receive only a few thousand dollars per course teach half of these classes. These faculty need help to become good mentors, get involved in educational research, and adopt good pedagogical practices, said Brown. In her response, Bragg noted that national statistics point toward something like 60 percent of the sections of precollege mathematics being taught by part-time faculty, and overall part-time faculty teach an estimated 45 percent of all mathematics two-year college sections.

Sally Johnson from the College of Southern Nevada said that her school gave 10,000 placement tests in the fall of 2011, and it provided students with options in taking the test. Nevertheless, 60 percent of those 10,000 students ended up in the lowest levels of mathematics, which are the equivalent of fifth grade and ninth grade mathematics. As she phrased it, "That is the reality of what we have on the ground." Furthermore, a student who starts in the fifth grade-level developmental mathematics class has approximately a 3 percent chance of ever taking a college-level mathematics class, she said.

Why are students enrolled in these classes when so many fail, asked Packard, commenting that "it is heartbreaking." If money is going to be invested in running so many sections of developmental mathematics, faculty also need development and support.

Carl Wieman, associate director for science in the White House Office of Science and Technology Policy, questioned the unusually high reliance on diagnostic tests and sorting in mathematics. In that respect, mathematics differs dramatically from other disciplines, which tend not to identify a lack of preparation as a deficiency. Biology, physics, and chemistry have courses for students who have not taken high-level classes in these subjects in high school. As an example of an alternative approach, George Boggs said that some colleges have been giving refresher courses before students take the assessment exam, and some of these students then do not have to go through a whole semester of developmental mathematics.

Jeannette Mowery from Madison Area Technical College, who was listening on the live webcast, e-mailed comments to the summit regarding developmental mathematics. She pointed out that, with few exceptions, mathematics is taught in isolation at all educational levels and not in context as a necessary tool to solve interesting and complex problems in a variety of industries and STEM application areas. All students would learn more mathematics if it were taught in context, she contended. She also pointed out that the level of mathematics needed for the majority of technical occupations is not higher mathematics such as trigonometry or calculus. Yet counselors and the standardized test system imply that students need to master mathematics at this high level to succeed in the sciences. "It is just not true, and it is a major barrier to students' success in the STEM field," she wrote.

Joan Sabourin from the American Chemical Society posed the challenge of decreasing the number of developmental mathematics and reading courses taught at two-year colleges by 5 percent each year through collaborations with K-12 institutions to increase the skills in mathematics and reading of 5 percent of K-12 students each year.

6

Transfer from Community Colleges to Four-Year Institutions

Important Points Made by the Speaker

- The numbers of community college students who transfer to four-year colleges and earn degrees in the natural sciences and engineering need to be greatly increased.
- Transfer scholarships focused on STEM fields and "individual development accounts" could help increase the diversity of students in STEM fields.
- Evidence-Based Innovation Consortia (EBICs) could create networks of relationships among community colleges, universities, and open education resource practitioners to support the adaptation and adoption of evidence-based innovations.

In analyzing transfer from community colleges to four-year institutions, Alicia Dowd, co-director of the Center for Urban Education and professor of higher education at the University of Southern California and author of one of the commissioned papers (see Appendix D), cited a recent report from the National Science Board (2010). That report called for providing quality science and mathematics teaching to all students, improving talent identification, and creating supportive ecosystems through professional development for STEM educators. All three steps are needed to enhance the flow of students from community colleges to four-year institutions, Dowd said.

EFFECTIVE TRANSFER POLICIES

Using survey data from NSF of recent college graduates, Dowd and her colleagues have been examining degree choice among Latino and Latina students who earn an associate's degree on their way to a bachelor's degree. They have found that the majority of students who transfer from a two-year college to a Hispanic-serving institution and earn a STEM degree do so in the social and behavioral sciences. Very few receive degrees in engineering, the physical sciences, or the biological, agricultural, or environment sciences.

The culture, values, and beliefs of faculty are critical factors contributing to the lack of transfer students in the natural sciences and engineering, said Dowd. Faculty members need to be partners in redesigning transfer systems, and they need robust evidence about what is effective and what is ineffective.

Transfer scholarships focused specifically on STEM fields could have a powerful effect on students and institutions, Dowd proposed. In addition, individual development accounts— savings accounts that are matched by public and private sources—could help increase the diversity of students in STEM fields.

Structural reforms of the curriculum, mentoring programs, and cultural transformation are all necessary, Dowd stated. Moreover, funds are available through recent federal initiatives to take appropriate and targeted action.

EVIDENCE-BASED INNOVATION CONSORTIA

Dowd suggested the creation of what she called Evidence-Based Innovation Consortia (EBICs). Their overall intent would be to facilitate transformational educational innovations that enable all students to thrive. By working with community colleges, universities, and open education resource practitioners, EBICs could create networks that would support the adoption and adaptation of evidence-based innovations. These networks would include agencies, organizations, industry, foundations, and others interested in specific topics, such as the reinvention of the mathematics curriculum. They would support the development of effective tools for systemic interventions to achieve educational performance and equity goals, such as equity scorecards. Finally, they would conduct and support research to gather and analyze evidence of innovations' effects.

Individual centers could focus on particular areas of innovation. For example, a center focused on the reinvention of the mathematics curriculum could coordinate the work of college faculty, researchers doing studies of curricula and pedagogy, and educators who are implementing innovative approaches.

These consortia would need to be prestigious, Dowd emphasized. For example, a high level of prestige among the EBICs could motivate faculty to participate more actively in improving transfer processes. The transfer rate for the most competitive private institutions has dropped from around 10 percent of student enrollments in 1990 to a little more than 5 percent in the most recent available data, Dowd noted. Other institutions enroll a higher percentage of transfer students, but the percentages at these institutions also have been declining. "That needs to change," said Dowd, because society "needs students who start out in community colleges and enter into the professions."

Collective Observations from a Breakout Group on Transfer

Individuals from the breakout group on transfer issues began the report to a plenary session of all participants by emphasizing articulation, alignment, and advising. Articulation agreements could benefit from greater clarity in terms of their scope, application to practice, and sustainability. Better alignment is needed between two- and four-year institutions, which will require that faculty members work together and collaborate on these issues. The members of the breakout session emphasized that "alignment" includes social and psychological components in addition to academic and institutional ones.

Second, improving transfer of students could benefit from research that explores the incentives and disincentives for effective transfer that can then drive changes in these incentives. For example, how might funding agencies promote incentives and diminish recognized barriers such as the cultural differences between institutions? State rankings of articulation and transfer policies based on research also might be a way to drive change.

Third, federal funding that allows students to have paid STEM-specific experiences—for example, through work-study or internship programs—could encourage more students to pursue STEM careers that would require successful transfer to four-year institutions.

DISCUSSION

Discussions on transfer issues during the summit centered on two broad issues—pathways and partnerships.

As opposed to the traditional image of a pipeline leading from K-12 education through college to graduate school and a career as a scientist or engineer, the concept of "pathways" is more appropriate for community colleges. Students can earn a variety of degrees and certificates from community colleges and either enter the workforce or a four-year institution. In many cases, students with four-year degrees return to community colleges to receive more specialized training.

Historically, students who enroll in technical education and receive applied associate degrees have transferred a limited amount of credits to four-year institutions, observed Debra Bragg. Since these students are some of the most diverse of all community college students, this poses an equity issue since technical students are unlikely to go on to earn a bachelor's degree. Bragg's research has looked at state policies that allow technical students to transfer to four-year institutions or enroll in a community college to earn an applied baccalaureate (AB) degree. Whether earned in a four- or two-year institution, AB degrees are spreading, and many of these arrangements are struck between particular two- and four-year colleges and not reflected in larger state policy. Bragg's research examines the various kinds of agreements that are associated with AB degrees, and she and her colleagues at the University of Illinois document and disseminate these degree arrangements.

Becky Packard observed that by the time community college students decide to become science or engineering transfer students, it may be too late, because they are already so far behind in taking the prerequisite courses that are needed to transfer. Students can become excited about physics and then get a reality check when they realize that they cannot major in the subject at a four-year school.

Advising and orientation sessions can be critical in keeping students' options open, especially if this guidance is specific to disciplines. For example, Jose Vicente of Miami Dade College noted that the college has launched a major program providing discipline-based orientation, not just in STEM fields but across the curriculum. Students benefit tremendously because they can see the roadmap for the entire period that they are at the institution. Also, in Florida, the legislature has allowed community colleges to provide baccalaureate degree programs for the past decade, and the students on this pathway are doing "stupendously," according to Vicente. The higher education system is even developing a roadmap for such students to go on to graduate school.

Eun-Woo Chang of Montgomery College in Maryland observed that a major weakness of counseling in community colleges is that few counselors are familiar with STEM majors. He suggested engaging more STEM faculty in the academic advising process to compensate for this weakness. Montgomery College, for example, has engineering faculty provide academic advising for engineering students.

John Morton from the University of Hawaii Community Colleges said that officials at community colleges in that state found that many students who were interested in transferring to a four-year institution in a STEM field were disadvantaged in pursuing an associate's degree compared with students at four-year institutions. As a result, the colleges instituted a transfer degree that is more parallel to bachelor's degree requirements. "Identifying [those students] as a cohort gave [them] an identity, and we have seen a big increase in the number of students pursuing that path to the baccalaureate," he commented.

Catherine Didion from the National Academy of Engineering and a co-principal investigator for the project emphasized the value of many two-year degrees in areas such as information technology and biotechnology. In promoting transfer policies, two-year pathways also need to be more transparent to students. Linnea Fletcher from Austin Community College agreed with this observation, pointing out that not every STEM occupation requires a four-year degree. High school counselors and institutions need more information about what is actually needed for particular jobs so that students have a more realistic idea of how much education is needed for those positions. Also, many STEM jobs are now and will continue to be in currently unanticipated fields, such as high-technology welding or fashion design. "This type of information is not being disseminated," said Fletcher.

Martin Storksdieck from the National Research Council and a co-principal investigator for this project pointed out that community colleges are not just educating future STEM professionals. They also are contributing substantially to the future scientific literacy of the general public, including K-12 teachers. The STEM courses that people take in community colleges are often their last formal courses in those subjects. "What does that mean for the way in which we want to structure them?" asked Storksdieck. He also observed that community colleges may need to examine their curricula and instruction as more students who enter these institutions take Advanced Placement and International Baccalaureate courses while in high school.

Geri Anderson from the Colorado Community College system raised the issue of the metric used to evaluate workforce training programs by the U.S. Department of Labor. Today the department is focused on getting people quickly into the workforce using high-demand certificates and degrees. She suggested that the Labor and Education Departments should engage in a dialogue about the value of education with longer term goals and the use of a different metric of success.

Articulation remains a problem for many institutions, responded Assistant Secretary Oates. Students should be able to transfer credits from two-year colleges to four-year colleges and have those credits count toward their major. This problem can be particularly acute in mathematics. In New Jersey, where she had worked previously, the rigor of calculus was not the same at community colleges as at four-year institutions. Faculty-to-faculty conversations are needed to harmonize the courses at the two types of institutions, she said. In New Jersey, those conversations not only educated community college faculty about what was needed but also helped open the eyes of four-year faculty about the talent at the community college level.

In response to a question from George Boggs about existing partnerships among two-year and four-year institutions, Dowd mentioned a conversation with a mathematics department chair at a comprehensive college who said that her institution's relationship with a nearby community college had deteriorated because they had lost funding for lunches that used to bring the two sets of faculty together. When a new grant enabled them to have lunch together again the relationship—and the transfer of students—improved. "It is good to have lunch," she said. "The realities of the structural alignment are going to be realized through human relationships."

Susan Elrod from Project Kaleidoscope at the Association of American Colleges and Universities pointed to the difficulty of forging robust partnerships among two-year and four-year institutions. Even with funding from the Gates Foundation to create such partnerships, it was difficult to figure out which people to bring together. "Getting the right people from the right institutions together in a room consistently to make sure that the messages are consistent, and to make sure that students feel no shift in culture, is really important," she observed.

Steve Slater from the Great Lakes Bioenergy Research Center at the University of Wisconsin–Madison pointed out that many universities do not view bringing in students from community colleges and helping them succeed as a high priority, which means that it is particularly hard to get junior, pre-tenured faculty involved in such efforts.

Tom Bailey from Teachers College, Columbia University, said that community colleges need coherent programs that span institutions if students are going to be able to transfer successfully. More than alignment is needed—the programs need to be coherent across institutions. Many students enter community colleges without much direction. They go into general studies programs, taking courses here and there. Some manage to earn a degree, though it may not be very coherent. As Bailey said, "We need to ask, what is it that we are doing to help students, [especially] if we have a particular interest in STEM?"

Bailey also observed that students who are already interested in STEM fields are the low-hanging fruit. Community colleges need to serve the needs of these students, but they also need to examine ways of getting more students engaged in STEM subjects.

Judy Miner from Foothill College observed that community colleges are critical components of the P-20 education continuum. They are uniquely positioned, have multiple missions, and feature open access for students. "The broad diversity of both our students and our institutions is not a problem to be solved but an opportunity to be seized, thereby empowering the most vulnerable of populations and in turn uplifting us all," said Miner.

7

General Discussion

During the discussion periods throughout the summit, participants considered many issues, beyond those covered in the previous three chapters, which have vital effects on community colleges. This chapter combines important points from those discussions into several broad topics. It also includes a box on major issues identified by the summit participants in a pre-summit survey described in Chapter 1.

SUPPORT FOR COMMUNITY COLLEGES

Several participants at the summit pointed to the immense financial challenges now facing community colleges. Community colleges "have really been hit," according to Jane Oates, and not just by the recent recession. Community colleges are often seen as having the capacity to raise funds in ways other than governmental support. As a result, policy makers and funding organizations view higher education—and community colleges in particular—as a lower priority than K-12 education, she said.

"We are not funding community colleges adequately," said Alicia Dowd. "We are not funding public education adequately." The counselor-to-student ratio at community colleges is at best 1 to 1,000, Dowd said. Adjunct faculty need better compensation and responsibility for fewer classes. Faculty need professional development and engagement with their peers as well as with students. "We need to pay for public education, including community colleges, and for faculty who are on campus," she said.

George Boggs agreed that if America is to meet the challenges of the future, policy makers must support colleges and universities as well as their students. States have cut funding to public higher education, including community colleges, despite a surge of enrollments. As a result, hundreds of thousands of students are being turned away because of inadequate resources. Boggs also observed that although part-time or adjunct faculty can do a great job in the classroom, a core of full-time faculty is essential to make policy changes and to work with colleagues at four-year institutions.

Mark Hubley from Prince George's Community College in Maryland pointed out that when he started at the community college in 2002 his department had 15 full-time faculty. Today, the enrollments in his department are twice what they were—and the department has 16 full-time faculty. The faculty in the department teach at four locations in the county, the college has a program on the campus that enrolls high school students, and the department teaches dual-enrollment classes at five high schools in the county. He said, "As excited as I get about things that we hear in meetings like this, it also makes me feel overwhelmed." In the future, said Hubley, the college will continue to need to do more with less, commenting that "anything you can do to help the faculty at community colleges will be most helpful."

Funding priorities can be a force for change. As Karl Pister, former chancellor of the University of California, Santa Cruz, and a member of the organizing committee observed, community colleges face the simultaneous challenge of educating potential transfer students, adults coming back to school, students taking developmental courses, and students interested in technical programs. In contrast, many four-year institutions have a "very monolithic culture" organized around conducting research. Research universities in their modern form were created when the federal government began making large investments in research in higher education following World War II. The same sort of change is required to spur major changes at the community college level, said Pister. "It is an application of the Golden Rule," he said. "People change their culture when the people with the gold change the rules." For example, federal agencies need to insist in their grant making that transfer be substantially increased. In Pister's view, "Without that kind of incentivizing, I don't think we are going to see much in the way of change."

Malvika Talwar from Northern Virginia Community College talked about the changes that targeted funding could make in the ranks of community college faculty. If community college careers were more palatable or exciting for students who are in graduate school, more PhDs would be interested in going this route. "When we are in graduate school, we don't really know much about community colleges," she said. One possibility would be postdoctoral teaching fellowships that are geared specifically toward community colleges. Such options would allow more students to explore that option and encourage mentors to talk about that route.

Support from policy makers and foundations is important, but the goals of improving educational attainment, particularly in STEM fields, will be met only if educators take responsibility for improving students' success, said Boggs. College and university faculty and administrators need to work together to improve completion rates and to facilitate the transfer of students from community colleges into upper division coursework.

Oates sounded a rare positive note regarding funding when she pointed out that the number of students with Pell grants who are in community colleges has gone up by more than 50 percent since 2009. She also said that more scholarship money is becoming available for students in both two-year and four-year institutions. The Obama administration has called for far more students to go to college and earn STEM degrees to help the economy grow. Revitalizing STEM education is not enough, Oates said. The challenge, she said, "is to embed STEM education as fundamental to America's future."

BUSINESS PARTNERSHIPS

Community colleges tend to have particularly close relationships with businesses and industries, for several reasons. Many offer career and technical education for occupations in nearby communities. In addition, employment is essential for many community college students. As Becky Packard pointed out, some students work full time and take one course at a time, though it may take them many years to earn a degree. Workplace tuition reimbursement programs can be particularly attractive options for such students.

Dowd suggested several other valuable roles that businesses could play. One is to fund transfer scholarships, which could raise the prestige of transfer students. Another would be to help establish and support community-based individual development accounts in which businesses would match money set aside for education. Celeste Carter from the National Science Foundation mentioned the possibility of businesses posing problems to groups of students that they could solve collaboratively. Packard added that linking scholarships with work-based internships could spur career development for students, rather than having jobs conflict with education.

Elaine Johnson from Bio-Link cited the importance of internships in steering students into STEM careers. Internships, particularly if they provide mentors and role models for students, can have a profound influence by showing students examples of success. "How do we create that environment of welcoming and getting over some of the fear that students have that they are not going to be successful in STEM careers?" she asked.

Bill Green from Accenture, a $25 billion company with 250,000 employees around the world, thanked everyone in the room for working on this issue. Solving these problems is not easy, he said, "but at the end of the day we are solving [these problems] for the competitiveness of our country and the standards of living of our citizens." He, too, emphasized the contributions business can make. "You are trying to solve this problem on your own," he said, while businesses are ready and able to help if they are challenged. For example, business provides $3.5 billion a year in philanthropy to education. "You can help us give it in a smarter, a more focused, and an evidence-based way," he said to other summit participants.

UNDERGRADUATE RESEARCH

One of the most effective ways to interest undergraduate students in STEM fields and keep them engaged is to get them involved in research. Several summit participants pointed to the special difficulties community college students have in doing research. It takes "a huge amount of time and effort to do that well," said Steve Slater. Even 40-year-old dislocated workers going back to school, said Oates, need an opportunity to whet their appetite for STEM careers with a taste of research.

Innovative community college faculty are thinking about how to use research to teach and inspire, noted Carter. She pointed to the Council on Undergraduate Research as the source of several useful publications on models that community college faculty can use to integrate undergraduate research into community colleges (e.g., Council on Undergraduate Research, 2009). Slater similarly pointed out that one way to expose students to research is to integrate research into the classroom. For example, the data being generated by DNA sequencing and other genomics applications offer endless opportunities to do original research, even at the high school level.

A related problem is ensuring that transfer students have as many opportunities to do undergraduate research as students who started in four-year institutions. Transfer students may not know professors as well or have social networks that can open the doors to research experiences. Transfer students may also need help with transportation, child care, and financial support to participate fully in research.

PARTNERSHIPS WITH HIGH SCHOOLS

Partnerships between high schools and community colleges also can be extremely valuable for students, high schools and colleges, and businesses. In particular, dual-enrollment programs between high schools and community colleges can pique student interest in college and help prepare them for higher education. Dual-enrollment programs also can encompass four-year institutions. For example, Dowd observed that Santa Ana College in California has been developing relationships with the local high school district and with the California State University system to make pathways among the institutions clear and intentional, with scholarships for students as an incentive to graduate from high school and follow a pathway through community college to a four-year institution.

Other promising approaches are early-college high schools that are STEM specific, fast-tracked baccalaureate degrees, and greater flexibility on the part of four-year institutions in accepting credits. In such partnerships, councils of high school, community college, and university faculty can look at the curriculum and identify gaps. However, some summit participants noted that more selective four-year institutions are less likely to become involved in such collaborative efforts.

Jose Vicente observed that faculty-to-faculty exchanges are needed not only between community colleges and universities but also between the school system and community colleges. Such exchanges provide a much better understanding for K-12 faculty and the school system of what the community college curriculum entails. At the same time, community college faculty can gain a much better understanding of the issues being addressed in the school system.

Carter said careful thinking needs to be devoted to where dual enrollment works and where it does not. Some students are intellectually and emotionally mature enough to handle it, but others are not. In addition, Linnea Fletcher pointed out that dual-enrollment programs are threatened financially in many states. In Texas, for example, if a high school student signs up for a dual-credit course and passes the community college exam, the student does not have to pay tuition, but the state is now considering ending that program because of funding problems.

Elaine Craft from Florence-Darlington Technical College in South Carolina pointed out that dual-credit courses can run into resistance from high school teachers who are defensive about AP courses. "You get a lot of pushback for trying to put dual credit in anything that they have AP credit for. They like those classes. They like those students, and they don't want any competition," she said, noting this is especially a problem in mathematics. Other participants countered that dual tracks in high school can work if they serve different purposes.

SUGGESTED RESEARCH

Several suggestions were made regarding research that could improve the contributions of community colleges in STEM education. Boggs said that more research is needed on attracting students, retaining them, and having them successfully complete programs at the certificate level, the associate level, and beyond. Research also is needed in closing achievement gaps and breaking down barriers within and between institutions.

Policy makers are asking why public funds should go to public institutions, only to have students fail, Boggs noted. However, performance-based funding mechanisms can be detrimental if they are not carefully thought out and based on evidence of what works. What kinds of funding can create the right incentives, he asked.

Deborah Boisvert from the Boston-Area Advanced Technological Education Connections suggested doing research in conjunction with the new tax grants being made available to promote stackable credentials (which are sequenced credentials that can move an individual along a career pathway or up a career ladder). She also pointed out that inquiry-oriented introductory courses rather than traditional lecture-based courses may be more effective at retaining STEM students early in their college years, and these courses also could be the subject of research.

Catherine Didion emphasized the lack of data in certain critical fields, which makes it difficult to determine how to improve success. "There are some real gaps of knowledge," she said. Dowd also pointed to the importance of data that would enable faculty and administrators to see exactly where students are being lost. For example, a project in California called the California Benchmarking project looked at cohorts of students starting at the earliest levels of developmental or pre-college mathematics. It then asked faculty to look at their syllabi and ask whether they are enabling the needed learning outcomes. If a student leaves one classroom with a passing grade and is unable to succeed at the next level, that is "a very powerful data point," Dowd said. That project led to a student equity and success tool that enables colleges to look at cohorts in a fine-grain manner along milestones and momentum points.

Packard said that community colleges need a capacity for institutional research, which is often lacking today. Transfer is a shared issue, so they should be able to partner with four-year institutions in this research. In this way, institutions can leverage their resources and conduct much more useful investigations than either institution could do on its own.

Rebecca Hartzler from the Carnegie Foundation for the Advancement of Teaching said that community colleges engage in considerable institutional innovation, but this work is rarely published. The Carnegie Foundation is trying to harness the innovation occurring in mathematics classrooms and create a community to come together around, in this case, developmental mathematics.

Martha Kanter observed that much has been learned over the past few decades, so one challenge is to systematize the things that work and apply them elsewhere. The First in the World Program has a competitive fund to get four-year and two-year schools to engage in research on what works and to disseminate the results across the country. She asked, "How can we corral [research results] into something bigger?"

FINAL REMARKS

The time is right for a bold and ambitious interagency initiative that could break down silos, spur innovation, disseminate and implement best practices and successful models, and cultivate experimentation, said Judy Miner. And, as Monica Bruning from Iowa State University pointed out, the responsiveness and adaptability of community colleges make them ideal partners in such an initiative. "I can't think of a better group than the community college educational system in our country to handle [these changes]," she said.

Boggs said that the summit's success would hinge on what happened after the participants left and went back to their day jobs. "I hope this is not just a one-time event," he said. Discussions need to continue with the goal of developing a comprehensive and coordinated "agenda for researchers, policy makers, educators, foundations, and business leaders to help us move ahead."

Responses to a Pre-Summit Survey: "Big Ideas" to Increase the Potential of Community Colleges for STEM Education and Careers

In the survey sent to registered participants before the National Academies' Summit on Community Colleges in an Evolving STEM Education Landscape, respondents were invited to contribute one big idea or insight they have about increasing the potential contributions of community colleges to STEM education and careers. Their contributions included the following:

1. Building and strengthening STEM pathways between two-year and four-year institutions. Examples include creating a three-year curriculum focused on transitions to baccalaureate institutions for students lacking college readiness, developing degree completion models that build on two-year programs, providing opportunities for two-year and four-year faculty collaboration, and exemplary articulation policies and practices.

2. Promoting an inquiry-based model of STEM instruction across two-year and four-year institutions. Examples include student coaching, hands-on labs, and teaching methodologies that teach STEM content in the context of employability skills.

3. Instituting specific curricular programs that have proved effective in retaining students in STEM education and careers. Programs mentioned include workforce education programs and biotechnology programs.

4. Requiring articulation agreements as a means for creating viable and affordable pathways to STEM careers. One example is aligning occupational STEM curricula with academic curricula. Another is the articulation of complete technician education programs.

5. Providing better support systems for students. Types of support mentioned included grouping students into cohorts and addressing social, cultural, financial, and personal issues.

6. Adopting, publicizing, and promoting STEM education as a community college priority. STEM education at the community college level could play a major role in teacher preparation and workforce development. It also can reach rural and remote communities. Making STEM education a community college priority could encourage the development of two-year and four-year institutional partnerships, emphasize the importance of community colleges in the evolving STEM education landscape, and focus federal funding on this issue.

7. Initiating uninterrupted federal funding for two-year and four-year STEM education programs. NSF and other funding agencies could help strengthen STEM education by funding programs that start in two-year colleges and provide a seamless transition into four-year institutions.

8. Strengthening K-12 STEM preparation and achievement. Efforts that could be taken to improve the academic preparation of students in STEM include strengthening the K-12 curriculum, having students take a test to determine college or job readiness prior to leaving high school, having dual-enrollment programs that allow for transfer of credits nationally, and recruiting students from STEM academies.

9. Increasing the capacity and competitiveness of community colleges to receive grants from NSF and other federal funding sources. Disadvantages for community colleges in applying for federal funds for program improvement include a shortage of faculty time to develop proposals and manage grant projects, inconsistent college administration support for grants, lack of grant-writing expertise, insufficient internal and external partnerships, and limited resources for institutions to learn how to submit proposals and manage awards.

10. Establishing professional communities to work on specific STEM education challenges. Particular challenges mentioned include developing curricula, leveraging technology, promoting faculty development, and recruiting more women and minorities into STEM education and careers.

References

American Association of Community Colleges. (2011). *Fast facts, 2011.* Washington, DC: Author.

Bettinger, E. (2010). To be or not to be: Major choices in budding scientists. In C.T. Clotfelter (Ed.), *American universities in a global market* (pp. 69-98). Chicago: University of Chicago Press.

College Board. (2011). *Trends in college pricing 2011.* New York: Author.

Council on Undergraduate Research. (2009). *Undergraduate research at community colleges.* Washington, DC: Author. Available: http://www.cur.org/urcc/ [June 25, 2012].

Frey, W.H. (2012). *America's diverse future: Initial glimpses at the U.S. child population from the 2010 census.* Washington, DC: Brookings Institution

Hrabowski, F.A., III, Maton, K.I., and Greif, G.L. (1998). *Beating the odds: Raising academically successful African American males.* New York: Oxford.

Hrabowski, F.A. III, Maton, K.I., Greene, M.L., and Greif, G.L. (2002). *Overcoming the Odds: Raising academically successful African American young women.* New York: Oxford.

Mullin, C.M., and Phillippe, K. (2011). *Fall 2011: Estimated headcount enrollment and Pell Grant trends.* Washington, DC: American Association of Community Colleges.

National Academy of Sciences, National Academy of Engineering, and Institute of Medicine. (2007). *Rising above the gathering storm: Energizing and employing America for a brighter economic future.* Washington, DC: The National Academies Press.

National Academy of Sciences, National Academy of Engineering, and Institute of Medicine. (2011). *Expanding underrepresented minority participation: America's science and technology talent at the crossroads.* Washington, DC: The National Academies Press.

National Research Council. (2010). *Exploring the intersection of science education and 21st century skills: A workshop summary.* M. Hilton, Rapporteur. Board on Science Education, Center for Education. Division of Behavioral and Social Sciences and Education. Washington, DC: The National Academies Press.

National Research Council. (2011). *Assessing the 21st century skills: Summary of a workshop.* J.A. Koenig, Rapporteur. Committee on the Assessment of 21st Century Skills, Board on Testing and Assessment. Division of Behavioral and Social Sciences and Education. Washington, DC: The National Academies Press.

National Science Board. (2010). *Preparing the next generation of STEM innovators: Identifying and developing our nation's human capital.* Arlington, VA: National Science Foundation.

National Science Foundation. (2011). *Women, minorities, and persons with disabilities in science and engineering: 2011.* (No. NSF 04-317.) Arlington, VA: Author.

Pryor, J.H., DeAngelo, L. Palucki Blake, L., Hurtado, S., and Tran, S. (2011). *The American freshman: National norms fall 2011.* Los Angeles: Higher Education Research Institute, University of California.

Schneider, M.S. (2011). *The hidden costs of community college.* Washington, DC: American Institutes for Research.

Shkodriani, G. (2004). *Seamless pipeline from two-year to four-year institutions for teacher training.* Denver, CO: Education Commission of the States.

Tsapogas, J. (2004). *The role of community colleges in the education of recent science and engineering graduates.* (InfoBrief 04-315.) Arlington, VA: Division of Science Resources Statistics, National Science Foundation.

White House (2011). *The White House summit on community colleges: Summit report.* Available: http://www.whitehouse.gov/sites/default/files/uploads/community_college_summit_report.pdf [June 25, 2012].

Appendix A

Summit Agenda[1]

National Research Council
Teacher Advisory Council, Board on Science Education,
Board on Higher Education and the Workforce, and
Board on Life Sciences

The National Academy of Engineering
Program Office

Carnegie Institution for Science
Carnegie Academy for Science Education

Community Colleges in the Evolving STEM Education Landscape
December 15, 2011
Carnegie Institution for Science
1530 P Street, NW
Washington, DC 20005

8:30   Registration (breakfast available in meeting room)

9:00   Welcome and Introductions

- *Jay Labov*, National Research Council
- *Toby Horn*, Carnegie Academy for Science Education
- *Barbara Olds*, National Science Foundation
- *George Boggs*, Chair of the Summit Steering Committee

---

[1]Links to video archives and PowerPoint presentations for many of the speakers at this Summit are available at http://nas-sites.org/communitycollegessummit/tentative-topics-and-agenda/.

9:30   Community College Opportunities and Challenges in STEM

- *Jane Oates*, U.S. Department of Labor

  - The need for a robust, diverse STEM workforce
  - The potential of community colleges to meet this need

- *Eric Bettinger*, Stanford University

  - To Be or Not to Be: Major Choices in Budding Scientists
  - Student Perspective on STEM pathways (video)

- Questions and General Discussion

10:30   Break

10:45   Key Issues in Realizing the Potential of Community Colleges

- *Becky Wai-Ling Packard*, Mount Holyoke—Outreach and Mentoring Programs
- *Debra Bragg*, University of Illinois—The Two-Year Mathematics Curriculum
- *Alicia Dowd*, University of Southern California—Transfer from Two-Year to Four-Year STEM
- Panel Discussion Among Paper Authors
- Questions and General Discussion

12:30- Working Lunch: Focus on underrepresented minorities in STEM
2:00

1:00   Keynote Address: Expanding Minority Participation in Undergraduate STEM

- *Harvey Fineberg*, President, Institute of Medicine
- *Freeman Hrabowski*, Chair, *Expanding Minority Participation: America's Science and Technology Talent at the Crossroads*
- General Discussion

2:00   Introduction to Breakout Discussions

2:15-   Breakout Discussions
3:00
- The Two-Year Mathematics Curriculum
- Outreach, Recruitment, and Mentoring in STEM
- Transfer from Two-Year to Four-Year STEM

Discussion Questions

- How does this issue (e.g., transfer) affect teacher quality in STEM?
- What are the next steps in this area to improve student pathways in STEM?
- What institutional, state, or federal policies are needed, including financial aid policies?
- What steps can you take now to improve student pathways?
- What further research is needed to guide policy and practice?

3:00   Merging of Groups by Topic

3:15   Break

3:30   Reports from Breakout Groups—*George Boggs*

4:00   Realizing the potential of community colleges in the evolving STEM education landscape: Next steps for practice, policy, and research

- Committee Members' Reflections
- *Martha J. Kanter*, Under Secretary of Education, U.S. Department of Education
- *V. Celeste Carter*, National Science Foundation
- General Discussion

5:00   Adjourn

Appendix B

Effective Outreach, Recruitment, and Mentoring into STEM Pathways: Strengthening Partnerships with Community Colleges

Becky Wai-Ling Packard
Professor and Co-Director,
Weissman Center for Leadership and the Liberal Arts
Mount Holyoke College

EXECUTIVE SUMMARY

The community college is the modal higher education entry point for students across the nation and is even more typical for first-generation, low-income, racial-ethnic minority, and nontraditional-age college students. Complex barriers can decrease the feasibility of pursuing four-year degrees in science, technology, engineering, and mathematics (STEM) fields via community college pathways. Exemplary outreach and recruitment efforts can help to increase the numbers of students who enter these critical pathways, while effective mentoring strategies can mitigate barriers and improve retention. Specifically:

1. *Outreach* efforts need to target students *and* their families so they can learn about the many career options within STEM fields as well as tri-level partnerships and pathways that link high schools, community colleges, and four-year institutions. Federal agencies and state departments of education should prioritize funding initiatives that exemplify sustainable, coordinated approaches to outreach. The National Science Foundation's Advancing Technology Education (ATE) Program portfolio provides many effective models for replication and expansion.

2. *Recruitment* is more effective when students can see the feasibility and relevance of completing a four-year STEM degree. States should expand dual-enrollment programs and make recommendations for college-preparatory math sequences; local policies should lift restrictions so college classes can count for both college and high school requirements. Governmental incentives should be directed toward industry partners who provide STEM-specific internships. Additionally, federal grants can incentivize the redesign of STEM courses at the introductory level.

3. *Several mentoring initiatives* improve student retention, including developmental bridge programs, science scholar programs, peer-led supplemental instruction, and undergraduate research experiences. Further research is needed into the design principles that make mentoring initiatives more effective and scalable. Funding agencies should require student mentoring plans in all relevant grant proposals, similar to the National Science Foundation's requirements for postdoctoral researchers. Grant programs should target innovations such as part-time summer research experiences for nontraditional students and future community college faculty mentoring programs. State higher education offices should include mentoring and retention plans as criteria for approving academic programs. Beyond formal programs, institutions need to expand informal mentoring strategies across their campuses.

In sum, efforts to recruit and retain community college students pursuing STEM transfer pathways should be coordinated, well designed, and sustained. Changes in educational policies and grant requirements can help to strengthen the impact of these investments. By helping institutional leaders in four-year institutions to recognize the benefits of collaborating with and learning from community colleges, we can further strengthen these critical partnerships.

INTRODUCTION

This paper focuses on effective outreach, recruitment, and mentoring strategies that can increase the number and diversity of students who use community college pathways to earn four-year degrees[1] in STEM. Many occupations in STEM fields now require a four-year degree; individuals who earn a bachelor's degree on average earn hundreds of thousands of dollars more during their careers and have access to a broader range of

---

[1]The importance of STEM associate's and certificate programs is recognized; however, these programs are not the focus of this paper.

career choices than high school diploma recipients do (Carnevale, Smith, and Strohl, 2010). Increasing STEM degree completion in the United States has been identified as an issue of national priority to boost global competitiveness (National Governors Association, 2011). The United States does not sufficiently tap the talents of the nation's students as evidenced by the underrepresentation of women, racial-ethnic minority, low-income, first-generation, and nontraditional-aged college students in many four-year STEM degree programs (Bailey and Alfonso, 2005; Espinosa, 2011; National Science Foundation, 2007), and the high percentage of international students in U.S. graduate STEM programs (National Science Board, 2008).

Yet, this is where challenge and opportunity meet. Community colleges attract students from all backgrounds, especially those underrepresented in STEM, by the hundreds of thousands. Indeed the community college is the most typical entry point into higher education today, representing about 50 percent of college students (Bailey and Alfonso, 2005; Engle and Tinto, 2008; U.S. Department of Education, 2006). However, community college pathways to four-year degrees are not as effective as they could be. Each year, over four billion dollars in grants and state allocations are lost when new, full-time community college students do not return for a second year of study (American Institutes for Research, 2011). Transfer rates from community college to four-year institutions are low overall, especially for low-income students of color (Bailey and Alfonso, 2005; Engle and Tinto, 2008; Packard et al., 2011; Reyes, 2011). In addition, few women pursue a STEM transfer pathway (Packard et al., 2011; Reyes, 2011). In order to strengthen pathways to a four-year degree for students from diverse backgrounds, it is critical to identify effective outreach and recruitment strategies to attract students as well as mentoring strategies to mitigate barriers and improve retention. Further, we need to identify levers of change to expand these practices across the nation.

CONCEPTUAL FRAMEWORK: THE FEASIBILITY OF STEM WITH A COMPLEX SOCIAL ECOLOGY

Students develop their college and career plans within a complex social ecology. Bronfenbrenner's ecological model helps us understand the influence of interconnected contexts on students' learning and career trajectories (Bronfenbrenner, 1979). Students develop within many spaces including the home, school, and workplace. The relationships among contexts also influence students, such as how strongly parents and teachers communicate. Indirect influencers, including access to transportation or availability of jobs, also persuade students, as do the broader political or economic contexts within which we all live, such as being in an economic downturn.

Applied to the pursuit of STEM majors in college, it is typical for students to be influenced by the guidance of a family member or a family friend (Kim and Schneider, 2005; Packard, Babineau, and Machado, 2012). When first-generation college students and low-income students turn to the home, however, they are not as apt to gain access to knowledge about college navigation, leads on internships, or a financial cushion if complications arise (Carter, 2006; Stanton-Salazar, 2011). For example, parents without college experience may not know that job training certificates rarely contribute toward an associate's degree, or that certain prerequisites are necessary to transfer from a community college into a four-year program.

In terms of home-school relationships, low-income parents may feel intimidated to approach high school math and science teachers, missing out on school-based support (Packard, Gagnon, and Moring-Parris, 2010). High school preparation is also an issue. In a study of 5,000 Latino community college students in Los Angeles, researchers found after three years of study, less than 9 percent of students were in a position to transfer to a four-year school, mostly due to the need for developmental courses (Hagedorn and Lester, 2006). Thousands of students each year enroll in developmental math, but only a few continue on to the transfer-based math courses and into a four-year STEM program (Hagedorn and DuBray, 2010). On the positive side, community college students generally report having positive experiences in STEM education with dedicated teachers and smaller class sizes that encourage students to continue their education (Patton, 2006).

Financial background is a major predictor of college entrance and persistence. Students from lower-income backgrounds are most likely to delay going to college, work many hours while attending college, attend school part-time, and limit their time on campus—all factors that predict degree noncompletion (Complete College America, 2011; Institute for Higher Education Policy, 2010). Low-income, first-generation college students are four times more likely to leave college during their first year than their peers are, and more than three times less likely to transfer to a four-year school in a six-year time frame (Engle and Tinto, 2008). Simply put, delays discourage students from transferring; students may elect shorter-term educational programs such as drafting instead of a longer-term career choice such as engineering, or leave STEM completely as a result (Packard and Babineau, 2009). What adds to the juggle: about 50 percent of students are employed at least part-time, and many are working full-time, while going to school (American Association of Community Colleges, 2007; Perna, 2010). Also, when students work many hours, they may not be able to avail themselves of academic resources such as office hours or study group sessions (Complete College America, 2011; Institute for Higher Education Policy, 2010), thereby decreasing access to opportunities designed to grow students' academic capacities and their sense of capability (Lotkowski, Robbins, and Noeth, 2004). Although access to a relevant job can positively contribute to STEM persistence, these opportunities are not as frequent as one might think (Packard et al., in press).

Finally, the pressure on community colleges to provide an effective *and* efficient mechanism for transfer has intensified in recent years despite serious resource constraints (Dowd, 2007). A study of 400 students from nine community colleges in Los Angeles found that ineffective advising, influenced in part by extremely high student-to-counselor ratios, led to a lack of student information about transfer requirements (Hagedorn, Cypers, and Lester, 2008). In many states, community colleges are not aligned with one another (Packard, Gagnon, and Senas, in press). For example, Biology 101 is not necessarily the same course at one community college as it is at another community college, leading students even in the same state to lose credits and time (Handel, 2007). Going further, articulation agreements between community colleges and four-year schools still require attention at the disciplinary level; it is still often the case that students will take a STEM prerequisite at a community college and later earn only general credits, rather than credit for a major requirement, upon transfer to a four-year institution (Packard et al., 2011; Reyes, 2011). In addition, women earn the majority of associate's degrees within community colleges, but only 5 percent earn degrees in STEM fields (Hardy and Katsinas, 2010), with community college men outnumbering their female peers in STEM majors at a ratio of three to one (Espinosa, 2011). Negative stereotypes about STEM careers are still prominent and can deter students; racial-ethnic minority students may especially question their fit or feel alienated from STEM fields (Aikenhead, 2001; Carlone and Johnson, 2007).

To conclude, feasibility is a major factor to consider when conceptualizing why we do not see more students pursuing or completing STEM majors via community colleges. Students may not have ready access to information about college pathways, while others question their fit within the STEM profession. A lack of academic preparation, particularly in math, may thwart the pursuit of STEM fields while others may find STEM degrees less feasible to pursue due to time and financial barriers. Additional delays are often experienced during the transfer to a four-year institution due to curricular misalignment or ineffective advising, leading some students to let go of their STEM majors or pursue a shorter-term program of study. Unfortunately, current programs designed to improve outreach, recruitment, and mentoring are not yet sustainable on a larger scale; not enough students are experiencing best practice in these domains. In the next sections, I discuss effective ways to improve outreach into STEM, highlighting the effective leadership that community colleges have taken in this domain. Second, I address effective recruitment into STEM by focusing on efforts to make STEM degrees feasible and relevant. Third, I review highly effective mentoring practices that improve retention in STEM within community college pathways to a four-year degree.

Barriers for Community College STEM Students

* Limited knowledge about college navigation
* Financial—both time and cost
* Academic preparation in math and science; need for developmental courses
* Misalignment of core courses across community colleges and four-year schools
* Delayed, inconsistent advising, orientation, and mentoring
* Constraints affecting the academic and social integration of working students
* Self-doubt regarding capabilities
* Cultural fit with professional identity or four-year institution
* Limited sustainability of programs designed to improve recruitment and retention

---

OUTREACH: BUILDING RELATIONSHIPS AND CULTIVATING INTEREST

Students and their families need greater access to information about college-going, feel invited into the college environment, and see compelling options in STEM fields. "Outreach" in this paper refers to an initiative designed to inform or invite students into STEM pathways. Outreach techniques include information campaigns, career days, and job shadowing (Kelly and Schneider, 2011). Effective role model selection improves outreach; interactions with role models with whom students can relate, such as alumni from their own communities or students just a step ahead of them in their education, discussing challenges they overcame, can be most effective at motivating students (Packard and Hudgings, 2002).

Outreach is a worthwhile, yet limited investment, as its primary aim is to inform or spark interest, not necessarily to facilitate enrollment or long-term persistence. Any one-time initiative that simply exposes a student to a range of interesting career options on a single day is not likely to have much of an impact. However, even short-term programs that integrate scientists into classrooms or where students engage in hands-on, authentic activities can grow student knowledge and interest in science careers (Laursen et al., 2007). More time-intensive programs, such as pre-college summer programs where students are exposed to laboratory experiences over the course of several weeks on a college campus, can contribute to increases in interest for learning science or pursuing a science career (Markowitz, 2004). Without ongoing support and continued learning experiences, however, students may move away from a STEM interest over time toward another field where they do gain support (Packard and Nguyen, 2003). Thus, outreach efforts that are more time-intensive or sustained over a period of time are recommended.

Consortia

Statewide and nationwide consortia, where clusters of programs are coordinated, organized, or offered in collaboration across a region, state, or nationwide, or across age and grade levels, can have more compelling effects than single programs operating in isolation. When programs are linked, students can more easily move forward through STEM pathways and continue to experience encouragement and engagement in a progressive manner, and resources can also be better allocated and utilized.

Two projects funded by the National Science Foundation's Building Capacity in Computing Program demonstrate effective statewide initiatives: Georgia Computes! (see http://gacomputes.cc.gatech.edu/) and Massachusetts's Commonwealth Alliance in Information Technology Education or CAITE (see http://www.caite.info/). In each project, efforts are coordinated statewide and across levels of education. In Georgia, they coordinated middle and high school summer camps, linked high schools with colleges, and facilitated collaboration between college students and graduate students using online tools. In addition, the consortium works with high school teachers through the Georgia Department of Education to offer workshops on new approaches to stimulate interest in computing education. In Massachusetts, CAITE is also an exemplary model that has organized a cross-alliance focus on community college transfer with regularized regional meetings. Further, a number of national-level organizations were recognized by the White House for working across the nation to promote effective outreach in STEM. For example, the National Girls Collaborative (see http://www.ngcproject.org/) maintains a database of organizations that support outreach to girls in science. For more effective outreach, grants should prioritize collaborative projects that coordinate outreach efforts across a state or within tri-level educational partnerships.

Embedding College Access Within Communities

Colleges can work to embed themselves within their local communities and high schools so that people in those communities see an open door to higher education. The more relationship-building there is between community colleges and four-year colleges and universities, the more successful we will be at helping students and families to navigate these four-year college pathways. Many community colleges already provide leadership on this front. For example, Austin Community College (ACC) has successfully embedded itself in the community by using multiple methods to increase the number of students who enter its doorways. Of particular interest is the College Connection Program, which won the Texas Higher Education Coordinating Board 2006 Star Award. Through this partnership, ACC works with high school seniors in 15 school districts to provide admission and enrollment services on their high school campuses. A stunning 6,400 high school seniors in Central Texas received an admissions acceptance letter to ACC with their high school diplomas. From 2003 to 2005, enrollments increased nearly 38 percent; the program also encouraged students to enroll in other colleges and universities (Texas Higher Education Coordinating Board, 2011). One recommendation is that all high schools should have community college and four-year partners and should be supported by state-level funding. Ultimately, students need to get the message repeatedly that applying to a community college goes hand in hand with the pursuit of a four-year college and university.

The focus of this paper is the pursuit of STEM; thus, it is important that messages about college access include a disciplinary message (for instance, start a science degree or science career at this community college). Having a scientist or college science students in the classroom, from both community colleges and four-year institutions, can communicate a powerful message on this front. The Boston Area Technological Education Connections (BATEC) has provided leadership in developing effective outreach protocols (Salaam, 2007).

Finally, math and science teachers in high schools are critical partners to community colleges and four-year schools. Improving science education can contribute to the number of students interested in studying science. Community college science and math teachers often select their careers because of a deep commitment to teaching; they can play an important role by recruiting and preparing new K-12 teachers who are interested in STEM education (Patton, 2006). A strong program that joins K-12 teachers and postsecondary educators at the four-year level is the Teachers Occidental Partnership in Science (TOPS) Program. Within this progressive program, high school educators have access to a number of resources, including cutting-edge instrumentation in the natural science fields, access to technology-based web programs that can be integrated into the K-12 curriculum, and support in professional development. Additionally, TOPS is also successful due to its alignment with statewide policies from the California Department of Education (Occidental College, n.d.).

The National Science Foundation's Advanced Technological Education (ATE) division has many exemplary programs for outreach efforts, and more specifically, the development of education modules to disseminate math and science instruction across the country. This program is the largest within the NSF to focus on community colleges; the numbers of students, families, and institutions positively affected are numerous (Advanced Technological Education Centers, 2011). One program, coordinated by the Museum of Science in Boston, is focused on improving educators' understanding of engineering, science, and technology by infusing engineering and technology concepts and skills into core introductory science and education courses in community colleges and four-year institutions (Cunningham and Lachapelle, 2011). Many other STEM education program elements in the ATE portfolio are also worthy of replication and expansion. Thus, I recommend the importance of sustaining this program and disseminating its models more broadly.

Summary of Author's Outreach Recommendations

- Select compelling role models who are step-ahead peers or alumni
- Coordinate outreach efforts across states and through national-level consortia
- Provide funding for outreach when efforts are organized across states, stakeholders, and levels of education
- Embed community colleges and four-year institutions into communities by working closely with high schools and sending united messages about access
- Invest in STEM teacher education as a form of outreach for teachers and students

---

RECRUITMENT: CREATING STEM PATHWAYS FROM HIGH SCHOOL INTO COLLEGE

Recruitment goes one step further than outreach; beyond sparking an interest or expanding career knowledge, the goal is to enroll students in their first course or to pursue a STEM major. In this section, I focus on creating feasible pathways for students to pursue STEM college degrees while they are still in high school. STEM degrees need to be both realistic and compelling to catalyze action. As mentioned, too many students arrive at community college without requisite math courses. Initiatives that increase the number of students arriving at college with college-level math in place increase the possibility of pursuing a STEM transfer pathway. In this section, I will focus my review on dual-enrollment programs and STEM-specific early college high schools[2] as well as the ways in which career-relevant internships and redesigned introductory courses can make a STEM major more compelling.

Prioritizing Pre-college Math

Dual enrollment, also called concurrent enrollment, refers to a practice in which high schools students take college-level classes, often on a college campus, while working toward their high school diploma. During the 2002-2003 school year, almost a decade ago, only 5 percent of all high school students, or 813,000 students, across the nation took college credit courses, with 77 percent using dual-enrollment partnerships between their high school and community college to do so (Golann and Hughes, 2008). That number has dramatically increased. The National Alliance of Concurrent Enrollment Partnerships (NACEP) is an organization that accredits concurrent enrollment programs, for seven years at a time, ensuring that the courses high school students take throughout the nation are college quality and optimal for college readiness. There are currently only 66 NACEP-accredited programs; most are two-year college partnerships, while 24 are four-year institution partnerships.

A dual-enrollment experience is more successful in predicting future college-going when students have had a more authentic college experience. Authenticity is enhanced by having class at the college campus and having classes in mixed groups of high school and college students (Edwards, Hughes, and Weisberg, 2011). Across the country, state-level policies exist that aim to support success in dual enrollment (Karp et al., 2005). For example, 18 states have a mandatory state policy on dual enrollment in which high schools must inform students of program opportunities and accept credit. Dual credit means that the credits earned are applicable toward high school and college requirements (Edwards and Hughes, 2011). However, at the local district level, variability exists in whether the high school will accept a college course for both a high school and college requirement. By lifting this restriction, high schools can play an important role in compressing the higher education timetable.

Not all students have access to dual-enrollment programs; students with high grade point averages have been prioritized over other students, and sometimes dual enrollment is reserved for career and technical education students (Karp and Hughes, 2008; Karp et al., 2008). However,

---

[2]Although scholar cohort programs and developmental bridge programs are also effective recruiting mechanisms, I choose to highlight these approaches in the mentoring section of this paper in order to describe a comprehensive range of effective mentoring initiatives together.

based on research within Florida and New York, expanding the eligibility requirements to students with even lower grade point averages is recommended because students are more motivated to persist in college as a result of gaining college credit while in high school (Karp et al., 2008). This important research also suggested designing dual-enrollment sequences, because students who took more than one dual-enrollment course observed greater benefits, and to offer dual-enrollment courses tuition-free to aid those economically disadvantaged. Completing a college-level math course not only opens doors to STEM degrees, but also predicts college persistence in general (Moore and Shulock, 2009). Thus, a recommended sequence that would lead students to pre-college math is recommended.

In a related movement, early college high schools (ECH) offer merged high school and college experiences on a compressed timetable. Recent research based on five years of this practice in which high schools are housed on a college campus cautiously suggested positive outcomes and the need to continue longitudinal work in this domain (Berger et al., 2009). In 2009, there were more than 200 early college high schools in 28 states, accounting for 50,000 students, with students from a low-income background (70%) and ethnic minorities (59%) accounting for the majority of participants (Jobs for the Future, 2009; Webb and Mayka, 2011). Almost all ECH graduates earn college credits across many fields of study. For example, a striking 95 percent of students at the Hidalgo Independent School District in Texas, a rural high-poverty district serving mostly Hispanic students, earned college credits while enrolled in an early college high school (Nodine, 2011). In addition, one in three ECHs is STEM-specific (North, 2011). In 2009 and 2010, close to 1,500 students graduated from over 30 STEM-themed early college high schools. Of these graduates, 25 percent earned an associate's degree while in high school, and two-thirds went on to a four-year college. One STEM-specific early college high school is Metro in Ohio, which has a partnership with Ohio State University and the Battelle Corporation. In the past two years, all Metro students were accepted to college, many students received college scholarships, and all were STEM-ready (North, 2011). A continuation of STEM-specific early college high schools, with continued research study, is warranted.

In sum, policy makers can find ways to make these programs more accessible and help credits transfer. For one, it needs to be a more common practice where college credit hours can fulfill state requirements for days and minutes needed for students' high school graduation (Jobs for the Future, 2006). State and local district policies need to lift restrictions so that college courses can count toward high school requirements as well as college credit. The practice of dual or concurrent enrollment can also be facilitated by making use of technology, such as online or blended course modules, where more students can gain access to a college-level math course (Lovett, Meyer, and Thille, 2008).

Improving Relevance

Even if the major is feasible to pursue, barriers still exist in the form of negative career stereotypes and perceptions. Improving the relevance of STEM careers can help. Because community college students are more apt to work while going to school, it is important to make these work experiences career-relevant, as seeing the relevance of their school learning to a career can positive motivate students (Packard, Babineau, and Machado, 2012: Packard et al., in press). Government incentives can be used to increase the number of companies providing STEM internships to students. In addition, introductory courses need to be redesigned to help students to see themselves within STEM pathways and to see the careers as more compelling to pursue. The use of interdisciplinary courses, service-learning, and society-relevant materials may be particularly promising for enrolling female and underrepresented racial-ethnic minority students (Chamany, Allen, and Tanner, 2008; Coyle, Jamieson, and Oakes, 2006). Federal grants should be used to incentivize colleges and universities to redesign these courses and to study their impact.

Summary of Author's Recruitment Recommendations

- Compress timetables for college completion using dual enrollment and STEM-specific early college high schools, thereby increasing feasibility of STEM college pathways
- Prioritize completion of a college-level math course through recommended sequences
- Expand eligibility for dual-enrollment programs to a wider range of students including students with lower grade point averages
- Lift restrictions so that college courses can count toward high school requirements and college credits
- Provide governmental incentives to companies to provide STEM internships
- Provide grant incentives for colleges and universities to redesign introductory courses using interdisciplinary and service-learning approaches so students can see themselves within STEM pathways

---

MENTORING: COMPREHENSIVE NETWORKS AT TRANSITION POINTS

Typically, mentoring is described as a term depicting a close one-to-one relationship, often formalized and intensive, where an older, more experienced person helps to encourage and guide a younger, less experienced person (Crisp and Cruz, 2009). However, models of mentoring have transformed in the past two decades such that it is now more widely accepted that mentoring can be obtained through various sources, including professional organizations, online systems, and even shorter-term relationships (Furman et al., 2006; Packard, 2003b). What we know from examining research and best practice is that many different kinds of mentoring relationships contribute to persistence in college and within STEM specifically. Students are more likely to persist in STEM when they experience a combination of (1) socioemotional mentoring functions, such as encouragement or role modeling, and (2) instrumental mentoring functions, including academic support, college navigation, and career coaching (Packard, 2004-2005). When students have multiple mentors from a variety of contexts across home, school, and the community, they are more likely to obtain a wider range of mentoring functions (Packard et al., 2009). A constellation mentoring strategy, or having a set of strategically assembled mentoring relationships from different sources that provide a range of mentoring functions along one's pathway, is recommended to promote persistence and career success.

In my work, I use a *functional* approach to studying and designing mentoring initiatives, which means one is focusing on the functions provided by a range of mentors than on any one formalized mentor (Packard et al., 2009). A functional approach to mentoring is helpful when considering the experiences of underrepresented groups as it is unlikely one can find a singular mentoring source to provide all functions needed (Packard, 2003b). In addition, a functional approach assumes a broad conception of mentoring, and is consistent with a constellation mentoring strategy. For example, peer sources of mentoring can sometimes be overlooked when using traditional conceptions of mentoring; however, peers are very effective at promoting a student's sense of belongingness and academic capability and thus are often an important part of a student's constellation (Ensher, Thomas, and Murphy, 2001; Packard et al., 2011.

A broader economic context also provides insights into additional support needed to access available mentoring. For example, if a student does not have access to transportation or cannot afford to give up four summer weeks of pay in order to participate in an 8-week summer mentoring program, then a particular mentoring initiative may miss out on the target students. Furthermore, students need to be able to reassemble their networks at transition points. Indeed, the mentoring that helps students to enter community college and select a STEM major may be different from the mentoring that helps students persist in a STEM major after transferring to a four-year school.

Several mentoring practices have been highlighted in the literature as improving student retention, including:

- Transition mentoring programs such as developmental bridge programs, college success courses, learning communities, and scholar cohort programs;
- Academic mentoring programs including peer-led supplemental instruction;
- Career-relevant mentoring programs including undergraduate research experiences and online career mentors (MentorNet).

These mentoring practices can each have a strong impact on diverse students pursuing community college STEM pathways to four-year degrees. Next, I will briefly describe some promising design principles for each of these approaches.

Transition Mentoring Programs

Transition mentoring programs leverage same-stage peer mentoring by establishing smaller cohorts of students. By featuring regular academic and counseling support for students, the transition is eased. However, important research has been conducted that examines different instantiations of these programs, highlighting important design principles. In the next paragraphs, I discuss developmental bridge programs, college success courses, learning communities, and scholar cohorts in further detail.

The idea behind a *developmental bridge* program is that enrolling in the bridge program will help students to accrue necessary academic and social skills before officially entering college so that they will be more apt to succeed. In a rigorously designed study of eight different developmental bridge programs in Texas, program students gained several benefits compared to control students, such as greater academic success in math and writing courses, as well as a greater likelihood to enroll in future writing and math courses (Wathington et al., 2011). It was clear that a combination of regular academic instruction in math and writing, college success advising, and academic plus social support from upper-level students each contributed positively. However, the authors suggested that some institutions do not find developmental bridge programs to be economically viable, so these elements are sometimes integrated into existing developmental courses or learning community approaches already used by the institution. Related research has discussed the impact of *college success courses* or extended orientation programs; researchers have been surprised at the positive impact on persistence above students who do not participate in such programs (Advisory Committee on Student Financial Aid, 2008; Berger et al., 2009).

An alternative to or complement to developmental bridge programs is a *first-year seminar* or *learning community*. Research focused on first-year seminars, which place introductory students into a small course section often with a common advising hour (Goodman and Pascarella, 2006), and learning communities, where students co-enroll in multiple introductory courses simultaneously (Weiss, Visher, and Wathington, 2010), shows that these approaches are associated with positive outcomes as well. However, further study will give us a better sense of the important elements. A recent, rigorously conducted research study suggested that a "basic" model of a learning community where students are co-enrolled in two or more classes together is not sufficient; instead, a more elaborated model involving interdisciplinary course planning by the professors of the two classes and special advising of the students was more effective (Weiss, Visher, and Wathington, 2010). Specifically, Kingsborough Community College organized co-enrollment of racially diverse cohorts in three classes: developmental English, an academic subject, and one credit of college orientation. Students also obtained counseling and a book voucher, while faculty received special professional development. In comparison to the control group, students felt more integrated and engaged, they passed more courses and earned more credits in their first semester, and slightly more program group members were still in college two years later. At Hillsborough Community College, a more basic approach was used in at least the first two semesters where students co-enrolled in a developmental reading course and a college success course. In the third semester, the faculty increased their collaboration and linked the co-enrolled courses more closely together. Positive impacts on the learning community students in contrast to the control students were only observed in the third semester, suggesting that simply co-enrolling students does not make a learning community effective.

Yet a third instantiation that also leverages same-stage cohorts for peer mentoring is found in *scholar cohort* programs. A scholar cohort is a group of students selected to be part of a distinguished cohort, such as a team of science scholars, because of stronger academic qualities or leadership potential. Typically, a scholar cohort will stay together for at least the first year of college. Science scholar cohorts have been studied and are shown to promote persistence, particularly among first-generation college students and low-income students (Myers, Brown, and Pavel, 2010), as well as promote graduate school attendance in STEM fields (DesJardins et al., 2010). A very effective university-housed program is the Meyerhoff Scholars Program at the University of Maryland at Baltimore County. According to research on this program using a number of comparative samples, the Meyerhoff Scholars were more apt than peer counterparts were to persist in a STEM major, to go to graduate school in STEM, and to earn better grades. In addition to the rigorous selection process and strong financial support of students, the use of study groups, a required summer bridge program, a shared residential location, and professional mentoring are credited as important factors (Maton, Hrabowski, and Schmitt, 2000; Stolle-McAllister, Sto. Domingo, and Carrillo, 2011).

Another incredibly powerful example is the Posse Scholar model (see http://www.possefoundation.org). Most Posse Scholars are students of color and first-generation college students. The quotation from the Posse website is compelling, "I would have made it in college if I had had my posse with me." Thus, by creating a posse and sending the posse to college together, the students are more likely to persist than any one member would have alone. The results have been remarkable, with 90 percent of Posse Scholars graduating from college, and most going on to graduate school. Selection is highly competitive, and the program prepares the scholars months in advance, through workshops provided by a special trainer. The selective four-year college provides full merit scholarships and an onsite mentor with whom they meet every other week, among other resources. Of particular interest is Brandeis, the first university with a Science Posse, funded by Posse and the Howard Hughes Medical Institute. Their selection process, too, is highly competitive (10 chosen from 1,600 applications in New York City). Science posse students attend an intensive two-week science bootcamp, enroll in a math and science course in their first year, and are placed in a research laboratory in their first semester. Although the program is still under study, the Brandeis coordinators reported it shows early signs of success; when science posse students thinks about leaving STEM, they do think twice because of what that means for their peers, and the program sees more frequent STEM majors within these cohorts as a result.

Academic Peer Mentoring

Academic mentoring is critical for student success. Many colleges and universities have tutoring programs, but they are often critiqued as being remedial and potentially stigmatizing for students seeking help. In contrast to tutoring, academic peer mentoring programs such as supplemental instruction (SI) or facilitated study groups (FSG) can encourage a culture of excellence in STEM through peer academic support for all. SI sessions are fundamentally very different from tutoring because peer instructors lead a reprise of the lecture, with prepared interactive materials designed to target misconceptions and trouble spots. The underlying premise of an FSG is similar to SI, but the overall climate of an FSG can feel more like a collaborative study session with the leader facilitating problem-solving exercises. Developed by Uri Treisman, the FSG approach helps to de-stigmatize help-seeking and instills habits of mind that are predictive of excellence (Treisman, 1992). In these models, students have leadership positions to grow into as they advance in the major, providing even more incentive, due to improved mastery and commitment to their majors (Lockie and Van Lanen, 2008). The programs are cost-effective and help to alleviate burden on faculty in office hours.

Supplemental instruction is an approach with a strong research foundation (Bronstein, 2008; Peterfreund et al., 2008; Preszler, 2006; Rath et al., 2007). University of Missouri–Kansas City's National Center for Supplemental Instruction recommends a method in which gateway courses, or ones that appear to hold people back from progressing in the major, are identified by analyzing course grades and withdrawals, as well as by using faculty nomination and student nomination. Students who attend SI or FSG sessions tend to do better in the gateway class, with positive results documented in introductory and upper-level courses, across science disciplines, with even greater benefits to racial-ethnic minority students (Rath et al., 2007). In summary, colleges should ensure that students in gateway courses not only have tutoring services, but also have SI or FSG sessions attached to them, as these provide insurance for retention and support for excellence for all students.

Career Mentoring Experiences

Numerous studies have documented the positive influence of undergraduate research experiences as a critical career-related mentoring experience (Gregerman, n.d.; Kim and Schneider, 2005; Seymour et al., 2004). Students gain a better sense of the field, grow their skills, and increase their commitment to STEM fields. Students can benefit from research opportunities in the summer or during the academic year. One recommendation is to fund more flexible research experiences for undergraduates, such as part-time summer research programs, so that nontraditional-aged students, students who need to continue working in another job, or students with families would be more apt to participate. In addition, online career mentoring systems such as MentorNet have been effective at sustaining STEM career interests by providing access to industry career professionals (Packard, 2003b). To expand these types of program, businesses can be incentivized to provide STEM mentors. A mentoring program could be developed to support graduate students who want to learn more about becoming a faculty member at a community college.

As for recommendations on mentoring, I offer many. Ultimately, further research needs to be conducted into the design principles for effective mentoring and how to bring these initiatives to scale. Training for mentors (Pfund et al., 2006) and students (Packard, 2003a) needs to be expanded to improve the efficacy of mentoring programs. Additionally, we need to infuse informal mentoring strategies into the daily activities of faculty and staff across campuses. Indeed, one of the strongest predictors of student engagement and persistence in STEM fields is the quality and type of interactions with faculty (Amelink and Creamer, 2010; American Society for Engineering Education, 2009; Kim and Sax, 2009; Ohland et al., 2008; Vogt, 2008). For example, the NSF-funded Engage in Engineering project (http://engageengineering.org) is working to strongly infuse faculty-student interaction into colleges of engineering across the country. Institutions also need to expand their institutional research capacities so that mentoring initiatives can be studied and linked to retention outcomes. Finally, state-level higher education offices that approve new academic programs should require mentoring and retention plans (Massachusetts Department of Higher Education, 2011).

Summary of Author's Mentoring Recommendations

- Federal agencies and other organizations should require student mentoring plans akin to the National Science Foundation's postdoctoral research requirements.
- State-level higher education offices should require mentoring and retention plans for new programs
- Institutions should invest in their institutional research offices in order to study the effectiveness of bridge programs, college success, and supplemental instruction on retention within disciplinary majors

---

CONCLUDING COMMENTS

We know that students and their families need access to information about STEM college programs and career opportunities. By forming partnerships across high schools, community colleges, and four-year institutions, and by coordinating outreach efforts, we can reach more students, grow knowledge, and spark interest. Beyond this, students need to find STEM pathways feasible to pursue and this will become more likely by expanding dual-enrollment programs that emphasize the completion of college-level math courses. Next, we need to expand effective mentoring initiatives in order to sustain the progress of students in these pathways. Several mentoring program types have been documented to show positive outcomes at the initial transition into college (developmental bridge, college success, and scholar cohort programs), within academic courses (supplemental instruction), and into the major (research experiences).

Because additional research is needed to continue to identify the design principles that improve effectiveness when programs are brought to scale, community colleges will need assistance to build capacity for institutional research efforts in order to contribute to the emerging knowledge base. It is also the case that institutional leaders need to understand the value of recruiting and retaining diverse students in STEM from community colleges to four-year institutions so that all students, especially the most underrepresented, see STEM fields as fields they want to pursue and places in which they will thrive. Although there is much work to do on this important issue, we can also see the many partners in this work who can contribute.

REFERENCES

Advanced Technological Education Centers. (2011). *Partners with industry for a new American workforce.* Available: http://atecenters.org/wp-content/themes/ate-centers/docs/pdf/ATE_centers_impact2011-spread_final.pdf [December 1, 2011].

Advisory Committee on Student Financial Aid. (2008). *Transition matters: Community college to bachelor's degree.* A Proceedings Report of the Advisory Committee on Student Financial Assistance. Available: http://www2.ed.gov/about/bdscomm/list/acsfa/transmattfullrpt.pdf [December 5, 2011].

Aikenhead, G.S. (2001). Students' ease in crossing cultural borders into school science. *Science Education, 85,* 180-188.

Amelink, C.T., and Creamer, E.G. (2010). Gender differences in elements of the undergraduate experience that influence satisfaction with the engineering major and the intent to pursue engineering as a career. *Journal of Engineering Education, 99*(1), 81-92.

American Association of Community Colleges. (2007). *About community colleges: Fast facts.* Available: http://www.aacc.nche.edu/AboutCC/Pages/fastfacts.aspx [December 1, 2011].

American Institutes for Research. (2011). *The hidden costs of community colleges.* Available: http://www.air.org/files/AIR_Hidden_Costs_of_Community_Colleges_Oct2011.pdf [December 1, 2011].

American Society for Engineering Education. (2009). *Creating a culture for scholarly and systematic innovation in engineering.* Available: http://www.asee.org/about-us/the-organization/advisory-committees/CCSSIE/CCSSIEE_Phase1Report_June2009.pdf [December 5, 2011].

Bailey, T.R., and Alfonso, M. (2005). *Paths to persistence: An analysis of research on program effectiveness at community colleges.* New Agenda Series, 6(1). Indianapolis, IN: Lumina Foundation for Education. Available: http://www.eric.ed.gov/PDFS/ED484239.pdf [December 5, 2011].

Berger, A.R., Cole, S., Duffy, H., Edwards, S., Knudson, J., and Kurki, A. (2009). *Fifth annual Early College High School Initiative evaluation synthesis report. Six years and counting: The ECHSI matures.* Washington, DC: American Institutes for Research. Available: http://www.eric.ed.gov/PDFS/ED514090.pdf [December 1, 2011].

Bronfenbrenner, U. (1979). *The ecology of human development: Experiments by nature and design.* Cambridge, MA: Harvard University Press.

Bronstein, S.B. (2008). Supplemental instruction: Supporting persistence in barrier courses. *Learning Assistance Review, 13*(1), 31-45.

Carlone, H.B., and Johnson, A. (2007). Understanding the science experiences of women of color: Science identity as an analytic lens. *Journal of Research in Science Teaching, 44*(8), 1,187-1,218.

Carnevale, A.P., Smith, N., and Strohl, J. (2010). *Help wanted: Projections of jobs and education requirements through 2018.* Washington, DC: Georgetown University, Center on Education and the Workforce. Available: http://www.eric.ed.gov/PDFS/ED524310.pdf [December 1, 2011].

Carter, D.F. (2006). Key issues in the persistence of underrepresented minority students. In E.P. John and M. Wilkerson (Eds.), *Reframing persistence research to improve academic success* (pp. 33-46). San Francisco, CA: Jossey-Bass.

Chamany, K., Allen, D., and Tanner, K. (2008). Making biology learning relevant to students: Integrating people, history, and context into college biology teaching. *CBE-Life Sciences Education, 7*(3), 267-278.

Complete College America. (2011, September). *Time is the enemy: The surprising truth about why today's college students aren't graduating… and what needs to change.* Available: http://www.completecollege.org/docs/Time_Is_the_Enemy.pdf [December 1, 2011].

Coyle, E.J., Jamieson, L.H., and Oakes, W.C. (2006). Integrating engineering education and community service: Themes for the future of engineering education. *Journal of Engineering Education, 95*(1), 7-11.

Crisp, G., and Cruz, I. (2009). Mentoring college students: A critical review of the literature between 1990 and 2007. *Research in Higher Education, 50*(6), 525-545.

Cunningham, C.M., and Lachapelle, C.P. (2011). *Research and evaluation results for the Engineering is Elementary project: An executive summary of the first six years.* Boston, MA: Museum of Science. Available: http://www.mos.org/eie/pdf/research/EiE_Executive_Summary_Mar2011.pdf [December 5, 2011].

DesJardins, S.L., McCall, B.P., Ott, M., and Kim, J. (2010). A quasi-experimental investigation of how the Gates Millennium Scholars Program is related to college students' time use and activities. *Educational Evaluation and Policy Analysis, 32*(4), 456-475.

Dowd, A.C. (2007). Community colleges as gateways and gatekeepers: Moving beyond the access "saga" toward outcome equity. *Harvard Educational Review, 77*(4), 407-418.

Edwards, L., and Hughes, K. (2011). *Dual enrollment for high school students.* New York: Columbia University, Community College Research Center, and Career Academy Support Network.

Edwards, L., Hughes, K.L., and Weisberg, A. (2011). *Different approaches to dual enrollment: Understanding program features and their implications.* New York: Columbia University, Community College Research Center, Institute on Education and the Economy. Available: http://ccrc.tc.columbia.edu/Publication.asp?UID=971 [December 1, 2011].

Engle, J., and Tinto, V. (2008). *Moving beyond access: College success for low-income, first-generation students.* Washington, DC: Pell Institute for the Study of Opportunity in Higher Education. Available: http://www.eric.ed.gov/PDFS/ED504448.pdf [December 1, 2011].

Ensher, E.A., Thomas, C., and Murphy, S.E. (2001). Comparison of traditional, step-ahead, and peer mentoring on protégés support, satisfaction, and perceptions of career success: A social exchange perspective. *Journal of Business and Psychology, 15*(3), 419-438.

Espinosa, L.L. (2011). Pipelines and pathways: Women of color in undergraduate STEM majors and the college experiences that contribute to persistence. *Harvard Educational Review, 81*(2), 209-240.

Furman, T., Gardella, J.A., Pagni, D.L., Puri, A., Schrader, C.B., and Tucker, S.A. (2006). *Mentoring for science, technology, and mathematics workforce development and lifelong productivity: Success along the k through grey continuum*. Available: http://www.unc.edu/opt-ed/events/mentoring_workshops/documents/PAESMEMwhitepaper.pdf [December 1, 2011].

Golann, J., and Hughes, K.L. (2008). *Dual-enrollment policies and practices: Earning college credit in California high schools. Lessons learned from the Concurrent Courses Initiative*. Community College Research Center, Teacher's College, Columbia University. San Francisco, CA: James Irvine Foundation. Available: http://www.eric.ed.gov/PDFS/ED506585.pdf [December 1, 2011].

Goodman, K., and Pascarella, E.T. (2006). First-year seminars increase persistence and retention: A summary of the evidence from how college affects students. *Peer Review, 8*(3), 26-28.

Gregerman, S.R. (n.d.) *The role of undergraduate research in student retention, academic engagement, and the pursuit of graduate education*. Available: http://www7.nationalacademies.org/bose/Gregerman_CommissionedPaper.pdf [December 1, 2011].

Hagedorn, L.S., and DuBray, D. (2010). Math and science success and nonsuccess: Journeys within the community college. *Journal of Women and Minorities in Science and Engineering, 16*(1), 31-50.

Hagedorn, L.S., and Lester, J. (2006). Hispanic community college students and the transfer game: Strikes, misses, and grand slam experiences. *Community College Journal of Research and Practice, 30*(10), 827-853.

Hagedorn, L.S., Cypers, S., and Lester, J. (2008). Looking in the review mirror: Factors affecting transfer for urban community college students. *Community College Journal of Research and Practice, 32*(9), 643-664.

Handel, S.J. (2007). Second chance not second class: A blueprint for four-year institutions interested in community college transfer students. *Change: Magazine of Higher Learning, 39*(5), 38-45.

Hardy, D.E., and Katsinas, S.G. (2010). Changing STEM associate's degree production in public associate's colleges from 1985 to 2005: Exploring institutional type, gender and field of study. *Journal of Women and Minorities in Science and Engineering, 16*(1), 7-32.

Institute for Higher Education Policy. (2010). *Cost perceptions and college-going for low-income students*. Research to Practice brief. Washington, DC: Author.

Jobs for the Future. (2006). *Smoothing the path: Changing state policies to support early college high school. Case studies from Georgia, Ohio, Texas, and Utah Early College High School Initiative*. Boston, MA: Author Available: http://www.jff.org/sites/default/files/smoothingexsum.pdf [December 1, 2011].

Jobs for the Future. (2009). *A portrait in numbers: Early college high school initiative*. Boston, MA: Author. Available: http://www.eric.ed.gov/PDFS/ED504745.pdf [December 1, 2011].

Karp, M.M., and Hughes, K.L. (2008). Study: Dual enrollment can benefit a broad range of students. *Techniques: Connecting Education and Careers, 83*(7), 14-17.

Karp, M.M., Bailey, T.R., Hughes, K.L., and Fermin, B.J. (2005). *Update to state dual enrollment policies: Addressing access and quality*. Washington, DC: U.S. Department of Education, Office of Vocational and Adult Education. Available: http://www2.ed.gov/about/offices/list/ovae/pi/cclo/cbtrans/statedualenrollment.pdf [December 1, 2011].

Karp, M.M., Calcagno, J.C., Hughes, K.L., Jeong, D.W., and Bailey, T. (2008). *Dual enrollment students in Florida and New York City: Postsecondary outcomes* (CCRC Brief No. 37). New York: Columbia University, Community College Research Center.

Kelly, A.P., and Schneider, M. (2011). *Filling in the blanks: How information can affect choice in higher education*. Washington, DC: American Enterprise Institute for Public Policy Research. Available: http://www.aei.org/files/2011/01/12/fillingintheblanks.pdf [December 1, 2011].

Kim, D.H., and Schneider, B. (2005). Social capital in action: Alignment of parental support in adolescents' transition to postsecondary education. *Social Forces, 84*(2), 1,181-1,206.

Kim, Y.K., and Sax, L.J. (2009). Student–faculty interaction in research universities: Differences by student gender, race, social class, and first-generation status. *Research in Higher Education, 50*(5), 437-459.

Laursen, S., Liston, C., Thiry, H., and Graf, J. (2007). What good is a scientist in the classroom? Participant outcomes and program design features for a short-duration science outreach intervention in k-12 classrooms. *CBE–Life Sciences Education, 6*(1), 49-64.

Lockie, N.M., and Van Lanen, R.J. (2008). Impact of the supplemental instruction experience on science SI leaders. *Journal of Developmental Education, 31*(3), 2-4.

Lotkowski, V.A., Robbins, S.B., and Noeth, R.J. (2004). *The role of academic and non-academic factors in improving college retention.* ACT policy report. Iowa City, IA: American College Testing. Available: http://www.eric.ed.gov/PDFS/ED485476.pdf [December 1, 2011].

Lovett, M., Meyer, O., and Thille, C. (2008). *The Open Learning Initiative: Measuring the effectiveness of the OLI statistics course in accelerating student learning.* Available: http://jime.open.ac.uk/article/2008-14/352 [December 1, 2011].

Markowitz, D.G. (2004). Evaluation of the long-term impact of a university high school summer science program on students' interest and perceived abilities in science. *Journal of Science Education and Technology, 13*(3), 395-407.

Massachusetts Department of Higher Education. (2011, June). *Final report from the working group on graduation and student success rates.* (Report No. BHE 11-09.) Available: http://www.mass.edu/currentinit/documents/Final%20Report%20from%20WG%20on%20Graduation%20and%20Student%20Success.pdf [December 1, 2011].

Maton, K.I., Hrabowski, F.A., and Schmitt, C.L. (2000). African American college students excelling in the sciences: College and postcollege outcomes in the Meyerhoff Scholars Program. *Journal of Research in Science Teaching, 37*(7), 629-654.

Moore, C., and Shulock, N. (2009). *Student progress toward degree completion: Lessons from the research literature.* Available: http://www.csus.edu/ihelp/PDFs/R_Student_Progress_Toward_Degree_Completion.pdf [December 1, 2011].

Myers, C.B., Brown, D.E., and Pavel, D. (2010). Increasing access to higher education among low-income students: The Washington State Achievers Program. *Journal of Education for Students Placed at Risk, 15*(4), 299-321.

National Governors Association. (2011). *Using community colleges to build a STEM-skilled workforce.* Issue brief. Washington, DC: NGA Center for Best Practices. Available: http://www.eric.ed.gov/PDFS/ED522079.pdf [December 1, 2011].

National Science Board. (2008). *Science and engineering indicators 2008.* Arlington, VA: National Science Foundation. Available: http://www.eric.ed.gov/PDFS/ED499643.pdf [December 1, 2011].

National Science Foundation. (2007). *Women, minorities, and persons with disabilities in science and engineering.* Arlington, VA: Author. Available: http://www.eric.ed.gov/PDFS/ED496396.pdf [December 1, 2011].

Nodine, T. (2011). *College success for all: How the Hildalgo Independent School District is adopting early college as a district-wide strategy.* Boston, MA: Jobs for the Future. Available: http://www.jff.org/sites/default/files/college_success_for_all.pdf [December 1, 2011].

North, C. (2011). *Designing STEM pathways through early college: Ohio's Metro Early College High School.* Boston, MA: Jobs for the Future. Available: http://www.jff.org/sites/default/files/ECDS_DesigningSTEMPathways_081511_0.pdf [December 1, 2011].

Occidental College. (n.d.) *What is TOPS?* Available: http://departments.oxy.edu/tops/NEW%20STUFF/TOPS%20Program%20information.htm [November 15, 2011].

Ohland, M.W., Sheppard, S., Lichtenstein, G., Eris, O., Chachra, D., and Layton, R.A. (2008). Persistence, engagement, and migration in engineering programs. *Journal of Engineering Education, 97*(3), 259-278.

Packard, B.W. (2003a). Student training promotes mentoring awareness and action. *Career Development Quarterly, 51*, 335-345.

Packard, B.W. (2003b). Web-based mentoring: Challenging traditional models to increase women's access. *Mentoring and Tutoring, 11*(1), 53-65.

Packard, B.W. (2004-2005). Mentoring and retention in college science: Reflections on the sophomore year. *Journal of College Student Retention: Research, Theory, and Practice, 6*, 289-300.

Packard, B.W., and Babineau, M.E. (2009). From drafter to engineer, doctor to nurse: An examination of career compromise as renegotiated by working class adults over time. *Journal of Career Development, 35*(3), 207-227.

Packard, B.W., and Hudgings, J.H. (2002). Expanding college women's perceptions of physicists' lives and work through interactions with a physics careers web site. *Journal of College Science Teaching, 32*(3), 164-170.

Packard, B.W., and Nguyen, D. (2003). Science career-related possible selves of adolescent girls: A longitudinal study. *Journal of Career Development, 29*(4), 251-263.

Packard, B.W., Kim, G.J., Sicley, M., and Piontkowski, S. (2009). Composition matters: Multi-context informal mentoring networks for low-income urban adolescent girls pursuing healthcare careers. *Mentoring and Tutoring, 17*(2), 187-200.

Packard, B.W., Gagnon, J.L., and Moring-Parris, R. (2010). Investing in the academic science preparation of CTE students: Challenges and possibilities. *Career and Technical Education Research, 35*(3), 137-156.

Packard, B.W., Gagnon, J.L., LaBelle, O., Jeffers, K., and Lynn, E. (2011). Women's experiences in the STEM community college transfer pathway. *Journal of Women and Minorities in Science and Engineering, 17*(2), 129-147.

Packard, B.W., Babineau, M.E., and Machado, H.M. (2012). Becoming job-ready: Collaborative future plans of Latina adolescent girls and their mothers in a low-income urban community. *Journal of Adolescent Research, 27*(1), 110-131.

Packard, B.W., Gagnon, J.L., and Senas, A. (in press). Avoiding unnecessary delays: Women and men navigating the community college transfer pathway in science, technical, engineering, and mathematics fields. *Community College Journal of Research and Practice.*

Packard, B.W., Leach, M., Ruiz, Y., Nelson, C., and DiCocco, H. (in press). School-to-work transitions of career and technical education graduates. *Career Development Quarterly.*

Patton, M. (2006). *Teaching by choice, cultivating exemplary community college STEM faculty.* Washington, DC: American Association of Community Colleges. Available: http://www.aacc.nche.edu/Resources/aaccprograms/Documents/stemfaculty.pdf [December 1, 2011].

Perna, L.W. (2010). Understanding the working college student. *Academe, 96*(4), 30-33.

Peterfreund, A.R., Rath, K.A., Xenos, S.P., and Bayliss, F. (2008). The impact of supplemental instruction on students in STEM courses: Results from San Francisco State University. *Journal of College Student Retention: Research, Theory and Practice, 9*(4), 487-503.

Pfund, C., Pribbenow, C.M., Branchaw, J., Lauffer, S.M., and Handelsman, J. (2006). The merits of training mentors. *Science, 311*(5,760), 473-474.

Preszler, R.W. (2006). Student- and teacher-centered learning in a supplemental learning biology course. *Bioscene: Journal of College Biology Teaching, 32*(2), 21-25.

Rath, K.A., Peterfreund, A.R., Xenos, S.P., Bayliss, F., and Carnal, N. (2007). Supplemental instruction in introductory biology I: Enhancing the performance and retention of underrepresented minority students. *CBE–Life Sciences Education, 6*(3), 203-216.

Reyes, M.E. (2011). Unique challenges for women of color in STEM transferring from community colleges. *Harvard Educational Review, 81*(2), 241-262.

Salaam, J. (2007). *Community college outreach toolkit.* Available: http://www.batec.org/download/outreachtoolkitforweb.pdf [December 1, 2011].

Seymour, E., Hunter, A.-B., Laursen, S., and DeAntoni, T. (2004). Establishing the benefits of research experiences for undergraduates: First findings from a three-year study. *Science Education, 88,* 493-594.

Stanton-Salazar, R.D. (2011). A social capital framework for the study of institutional agents and their role in the empowerment of low-status students and youth. *Youth and Society, 43*(3), 1,066-1,109.

Stolle-McAllister, K., Sto. Domingo, M.R., and Carrillo, A. (2011). The Meyerhoff Way: How the Meyerhoff scholarship program helps black students succeed in the sciences. *Journal of Science Education and Technology, 20,* 5-16.

Texas Higher Education Coordinating Board. (2011). *2006 winners and finalists presented by the Texas Higher Education Coordinating Board.* Available: http://www.thecb.state.tx.us/index.cfm?objectid=411CBE31-0177-0788-C4A87D4C64D90C39 [December 1, 2011].

Treisman, U. (1992). Studying students studying calculus: A look at the lives of minority mathematics students in college. *College Mathematics Journal, 23*(5), 362-372.

U.S. Department of Education. (2006). *Profile of undergraduates in U.S. postsecondary education institutions, 2003-04: With a special analysis of community college students* (NCES 2006-184). Washington, DC: National Center for Education Statistics. Available: http://www.eric.ed.gov/PDFS/ED491908.pdf [December 1, 2011].

Vogt, C.M. (2008). Faculty as a critical juncture in student retention and performance in engineering programs. *Journal of Engineering Education,* (97), 27-36.

Wathington, H.D., Barnett, E.A., Weissman, E., Teres, J., Pretlow, J., and Nakanishi, A. (2011). *Getting ready for college: An implementation and early impacts study of eight Texas developmental summer bridge programs.* Available: http://www.postsecondaryresearch.org/i/a/document/DSBReport.pdf [December 1, 2011].

Webb, M., and Mayka, L. (2011). *Unconventional wisdom: A profile of the graduates of early college high school.* Boston, MA: Jobs for the Future. Available: http://www.eric.ed.gov/PDFS/ED519999.pdf [December 1, 2011].

Weiss, M.J., Visher, M.G., and Wathington, H. (2010). *Learning communities for students in developmental reading: An impact study at Hillsborough Community College.* NCPR brief. New York: National Center for Postsecondary Research. Available: http://www.eric.ed.gov/PDFS/ED512710.pdf [December 1, 2011].

Appendix C

Two-Year College Mathematics and Student Progression in STEM Programs of Study

Debra D. Bragg
Professor and Director,
Office of Community College Research and Leadership
University of Illinois at Urbana-Champaign

EXECUTIVE SUMMARY

In spite of the strident pursuit of standards-based reform of two-year college mathematics, implementation of reform has been slow and uneven. National studies show student enrollment in two-year college mathematics is growing in proportion to overall enrollment growth in higher education, but a substantial portion of these students are taking pre-college mathematics courses. Research suggests many of these students never reach college-level mathematics. Improvements need to be made to two-year college mathematics to prepare more students for science, technology, engineering, and mathematics (STEM)-related careers. Specific recommendations to support this goal are

Take a P-20 approach to reforming the entire mathematics curriculum. Without a strategic, collaborative endeavor, it will be difficult for two-year colleges that are caught between K-12 education and higher education to implement and sustain meaningful change.

Conduct more research on the teaching and learning of two-year college mathematics. Finding ways to support two-year college faculty to engage in professional development that reinforces innovative pedagogies is important. Included in this list are topics linked to quantitative literacy, accelerated and contextualized instruction, and college placement and related assessments that need to be better linked closely to student learning.

Investigate more fully the characteristics, experiences, and aspira-

tions of students who enroll in two-year college mathematics. More information is needed about how diverse learners, especially women and minorities, experience their two-year mathematics courses (pre-college and college level) and how these experiences influence their subsequent enrollment, completion, and career decisions related to STEM.

Engage practitioners in action research on mathematics education to facilitate the adoption and scale-up of innovation. Two-year faculty would benefit from opportunities to engage in action research that helps them to understand how various pedagogical and assessment strategies impact the learning of diverse students, and then employ these strategies in their classrooms.

INTRODUCTION

There is wide consensus that mastery of mathematics is essential to progressing into and through STEM programs of study, yet many students are unsuccessful at navigating the normative mathematics course sequence (Cullinane and Treisman, 2010) that is fundamental to their advancement into STEM-related careers. Recent concerns about international competition and the struggling economy have focused attention on this important issue and renewed concerns about the challenges that many students, particularly women and minorities, face succeeding in mathematics coursework (National Academy of Sciences, National Academy of Engineering, and Institute of Medicine, 2010). Resolving this problem is an urgent priority if the nation is to see growth in student enrollment and success in STEM programs of study, placement of graduates in STEM-related careers, and rejuvenation of the nation's economy.

This paper examines the influence of the two-year mathematics curriculum on students' progression into and through STEM programs by drawing upon extant literature, materials on the Internet, and personal communication with two-year college mathematics experts and practitioners. It acknowledges the expansive developmental mathematics curriculum offered by two-year colleges, but even more importantly, provides insights into college-level mathematics that has been overshadowed by a preoccupation with developmental education. The paper begins with a brief historical perspective and then proceeds to address such questions as: what is the status of two-year mathematics courses, who teaches them, and how are they taught? What standards-based reforms are associated with two-year college mathematics, what curricular and pedagogical innovations are capturing the attention of mathematics reformers, and what do we know about the impact of these reforms on student success? This paper concludes with recommendations for future research, policy, and practice on two-year college mathematics that is intended to enhance student progression through STEM programs of study and into STEM-related careers.

PERSPECTIVES ON MATHEMATICS CURRICULUM IN THE TWO-YEAR COLLEGE

A useful framework for understanding two-year mathematics curriculum comes from Cullinane and Treisman (2010), who label the mathematics curriculum in the United States the "normative mathematics course sequence" (pp. 7-8), which they claim is ubiquitous to the P-20 (primary through grade 20) education system. The normative mathematics course sequence extends from basic arithmetic, to pre-algebra, algebra, and intermediate algebra on to trigonometry, pre-calculus, calculus, and other calculus-based courses, with a fuzzy demarcation between pre-college and college-level mathematics that starts with college algebra. Geometry may be part of the sequential mathematics continuum, or it may be omitted, to the detriment of students' advancement into calculus and calculus-based sciences such as physics. Because this framework represents the dominant schema for which mathematics is taught and for which student competence is assessed at the secondary and postsecondary levels, I use this framework as the basis for discussing the literature. Later, in my discussion of reforms of the two-year college mathematics curriculum, I again cite Cullinane and Treisman (2010) who are studying alternatives to the normative mathematics course sequence. First, however, I provide a brief historical foundation and then move to contemporary developments in two-year college mathematics.

Liberal arts and sciences courses, including mathematics courses, have been part of the two-year college curriculum since creation of junior colleges in the early 1900s. Cohen and Brawer (1982) observed that, by the time two-year colleges arrived on the U.S. higher education scene, the academic disciplines were already "codified" (p. 284) by the rest of the educational system. Junior colleges that emerged to fill the void between high schools and universities adopted the prevailing curriculum structure advocated by the mathematics discipline and were therefore caught in between the K-12 sector and the four-year college sector from the start. To this end, Cohen and Brawer observed that, "the liberal arts [courses of two-year colleges] were captives of the disciplines; the disciplines dictated the structure of the courses; [and] the courses encompassed the collegiate function" (1982, p. 285). To facilitate the acceptance of college credits at the university level, two-year colleges reproduced the curriculum as well as the pedagogical methods used by universities to which their students sought entry.

Transfer was born from these early replication efforts. A landmark study of junior colleges conducted at mid-20th century by Medsker (1960) confirmed the lengths to which two-year colleges mimicked university curriculum to enhance students' ability to transfer. He noted, "the junior college forfeits its identity and its opportunity to experiment in the development of a program most appropriate for it" (p. 53). Looking back to the start of the comprehensive curriculum of the two-year college, Cohen and Brawer (1982) cited findings from a very early study of 58 junior colleges conducted in 1921 and 1922 by Koos (1924, p. 29) that showed liberal arts, sciences, and humanities courses dominated the early junior college curriculum, with three-fourths of all courses representing these disciplines. Across a broad array of the liberal arts and sciences, mathematics represented about 8% of all course offerings. Whereas mathematics was not as dominant as English, communications, and the sciences, it was nearly universally offered in the two-year college. By the late-1950s, a national survey conducted by Medsker (1960) of 230 two-year colleges in 15 states confirmed mathematics courses were ubiquitous to the two-year college curriculum, but still, only a relatively modest proportion of students enrolled in them. In fact, only about one-quarter of two-year colleges required their students to take at least one mathematics course to meet general education requirements. Medsker's study was also important because it was one of the first to document the prevalence of pre-college courses in reading, writing, and mathematics, foreshadowing a phenomenon that would continue to grow to the present day.

Several decades subsequent to Medsker's study, Cohen and Brawer (1987) studied the two-year college curriculum and found remarkably similar findings about mathematics course-taking. Their analysis showed 9 percent of total course enrollments in the liberal arts, sciences, and humanities curriculum were in mathematics, and again reflective of Medsker's results, the survey revealed a high proportion of mathematics courses were at the pre-college[1] level. Subsequent studies conducted by Cohen, Brawer, and colleagues included a curriculum mapping study conducted by Cohen and Ignash (1992) about two decades ago. This study examined courses offered by a national sample of two-year colleges by scouring the spring 1991 catalogs and course schedules of 164 community colleges, balanced by small (less than 1,500 students), medium, and large (over 6,000 students) institution size. Cohen and Ignash mapped the liberal arts and nonliberal arts curriculum into broad subject areas of which mathematics and computer science were combined into one area. Their study showed the preponderance of mathematics enrollments were

---

[1]Consistent with other literature on two-year college mathematics (see, for example, Blair, 2006), I use the term pre-college to refer to mathematics courses offered below the college level, including courses often referred to as developmental and remedial education courses.

in classes offered at the introductory and intermediate course levels, with enrollments at the introductory or intermediate level being nearly 9 times larger than enrollments at the advanced level (about 766,000 enrollments in the former and approximately 87,000 enrollments in the latter). Though enrollments were substantially lower in advanced mathematics courses, this study confirmed that two-year colleges offered a substantial array of mathematics courses, including courses extending from the developmental level to calculus, as well as statistics. The number of sections of mathematics accounted for 10.7 percent of the total liberal arts curriculum, which ranked mathematics just behind humanities at 13.4 percent and English at 12.8 percent.

Another important aspect of two-year college mathematics curriculum that is evident in the curriculum mapping study of Cohen and Ignash (1992), and that also has relevance to this discussion, pertains to the rise of nonliberal arts curriculum, a trend that began in the 1970s (Cohen and Brawer, 1987). Since much of mathematics course-taking in the two-year colleges relates to the majors that students choose in nonmathematics subjects, it is important to understand the ways mathematics is used to fulfill general education requirements. Cohen and Ignash (1994) identified the emergence of occupational-technical fields of study (many having a technical focus and having some STEM-related content) beginning in the 1970s, and they documented the growth of technical education, trades and industrial education, and other programs of study offered by two-year colleges that require various levels and forms of mathematics. Whereas the offering of liberal arts and sciences courses has been relatively robust over the years, by the 1990s nonliberal arts and sciences courses accounted for about 45 percent of the two-year college curriculum and occupational-technical education course were prominent among them. Technical mathematics and courses tailored for other majors such as elementary education were apparent in the curriculum as well. One implication of this trend is that the teaching of mathematics, which had been the purview of the mathematics discipline, spread to other instructional units and efforts to integrate mathematics with other subjects emerged as a strategy to increase learners' abilities to apply mathematics in diverse occupational settings (Grubb, 1999).

Concerns about students' lack of preparation for college-level mathematics were also growing during the latter decades of the 20th century, as noted by an American Mathematical Association of Two-Year Colleges (AMATYC) report that showed developmental mathematics courses were offered by 91 percent of two-year colleges (Baldwin and the Developmental Mathematics Committee, 1975). Literature documenting the growth in remedial enrollments in two-year colleges observed pre-college mathematics courses were necessary for "marginally literate students emanating from the secondary schools" (Cohen, 1984, p. 1). In a national survey conducted in the early 2000s, Greene and Forster (2003) found only 32 percent of all high school graduates demonstrated the level of competence needed to enter college mathematics coursework. Among all learners, Greene and Forster identified Hispanics and African Americans as "seriously underrepresented in the pool of minimally qualified college applicants" (p. 3), and they attributed their lack of preparation to inadequate K-12 education rather than "financial aid or affirmative action policies" (p. 3). Their research points to the uniquely important role that community colleges play in transferring underserved students to the university level in STEM-related careers (Arbona and Nora, 2009). However, transfer is not possible if students are unable to navigate the mathematics curriculum that begins at the pre-college level. Attewell et al. (2006) found 60 percent of community college entrants are required to take one or more developmental courses, usually mathematics, and numerous studies by Bailey and colleagues at the Community College Research Center (CCRC) (see, for example, Bailey, Jeong, and Cho, 2010) reveal the dismal success rates of students whose placement test scores prescribe multiple developmental mathematics courses. Beyond the developmental level, Adelman (2004) noted failure and withdrawal rates of 50 percent or higher in college algebra and pre-calculus courses, demonstrating problems with student success extend beyond to the gatekeeper course level. This trend and other critical aspects of the two-year mathematics curriculum, including documenting enrollment in mathematics course sequences extending from pre-college to college level, are examined in the next section.

CONTEMPORARY TWO-YEAR COLLEGE MATHEMATICS

Most of what we know about two-year college mathematics in the United States comes from a few large-scale national surveys. The fullest depiction of the curricular landscape about two-year college mathematics is the national inventory of the mathematics curriculum in U.S. higher education that has been conducted every five years starting in 1965 by the Conference Board of Mathematical Sciences (CBMS) with support from the National Science Foundation (NSF). *The CBMS2005: Fall 2005 Statistical Abstract of Undergraduate Programs in the Mathematical Sciences in the United States* is the last published installment of the national survey results on the two-year college mathematics curriculum (Lutzer et al., 2007). However, preliminary results of the 2010 CBMS national inventory on two-year college mathematics were shared with me by Ellen Kirkman and Rikki Blair (personal communication, December 4, 2011) to update this paper.[2] Taken together, the 2005 and 2010 reports (as well as selective use of earlier CBMS surveys) provide the most detailed description of two-year college mathematics curriculum offered by public community colleges in the United States, including trends in student enrollments, courses, faculty, and instructional practices.

The CBMS survey used a stratified, simple random sampling design, with strata based on the three variables of curriculum, highest degree level offered, and total institutional enrollments to address three distinct universes: two-year college mathematics programs, mathematics departments in four-year colleges and universities, and statistics departments in four-year colleges and universities. The stratum specifications used in the 2005 CBMS administration exactly replicated the ones used in the CBMS 2000, and closely emulates the specifications of previous CBMS surveys that were adjusted over time to improve national estimates. With respect to the 2005 CBMS administration, the most recent date for which a full report of methodology is available, a total of 600 public two- and four-year colleges and universities were surveyed during the period of September 2005 to May 2006. Minor adjustments are made to the CBMS at each administration, but the core of items included on the survey remains constant to address enrollments, instructional strategies, faculty demographics, and so forth.

Figure C-1 reveals fall enrollments in mathematics and statistics at a 5-year interval from 1975 to 2005 (Lutzer et al., 2007), compared to total enrollments in public four-year and two-year colleges obtained from the *Community Colleges, Special Supplement to the Condition of Education, 2008* report (Provasnik and Planty, 2008). Results suggest the enrollments in mathematics and statistics are growing commensurate with the increase in enrollments in public four-year and two-year colleges over the last 25-year period. Total enrollment growth in public two-year college is highly correlated ($r = .96$) with mathematics and statistics enrollment over this time span. More recent enrollment figures for Fall 2010 for mathematics and statistics show enrollment climbed to an all-time high of 2,096,000 (E. Kirkman and R. Blair, personal communication), mirroring the enrollment growth in public two-year colleges to an unprecedented high of 7,101,000 for 2009, the latest year statistics are available from the U.S. Census Bureau (2012). With respect to the growth in mathematics enrollment, the CBMS 2010 survey showed a 26 percent increase from 2005, which is about the same percentage increase that was observed between 2000 and 2005.

---

[2]I want to express my sincere appreciation to Ellen Kirkman and Rikki Blair for their generosity in sharing preliminary tables from the forthcoming CBMS2010: *Fall 2010 Statistical Abstract of Undergraduate Programs in the Mathematical Sciences in the United States*.

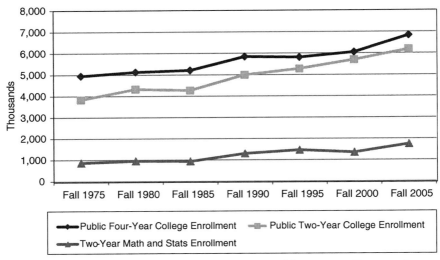

FIGURE C-1 Total public enrollment in two-year and four-year colleges and two-year mathematics and statistics enrollment (Fall 1975–Fall 2005).

Figure C-2 shows two-year college mathematics enrollments are critical to the overall U.S. postsecondary education system. Whereas two-year mathematics enrollment has seen some fluctuation over the 15-year period from 1990 to 2005, the 2005 mathematics enrollment figure confirms substantial growth from earlier years to the point where there were only modestly fewer enrollments in mathematics at the public two-year colleges (n = 1,580,000) than four-year public and private colleges (n = 1,607,000) by Fall 2005, based on 2005 CBMS results (Lutzer et al., 2007). These totals take into account dual enrollment, which has grown substantially over the last decade (Waits, Setzer, and Lewis, 2005); however, they do not take into account mathematics courses taught outside of mathematics disciplinary units, including centralized pre-college education units that are responsible for teaching pre-college mathematics classes. Therefore, these figures almost certainly underestimate enrollments in pre-college mathematics (arithmetic, pre-algebra, elementary algebra, intermediate algebra, and geometry) and possibly other mathematics-related courses taught on two-year college campuses, suggesting the actual enrollment in two-year college mathematics may be higher still.

Figure C-3 shows the percentage enrollment in two-year college mathematics courses by type of course and by the year the survey data were collected, based on the most recent CBMS 2010 data supplied by Kirkman and Blair (personal communication). Looking at the overall curriculum delivered by two-year college mathematics units, the preponder-

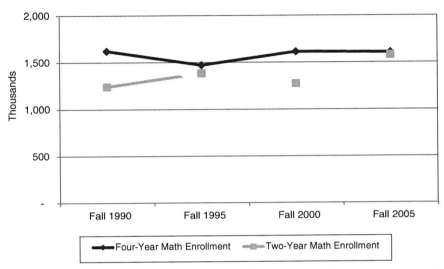

FIGURE C-2 Total enrollment in four-year college mathematics and two-year college mathematics (Fall 1990–Fall 2005).

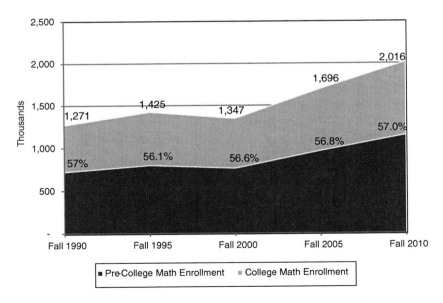

FIGURE C-3 Total two-year college enrollment in mathematics and percentage of total enrollment at the pre-college level (Fall 1990–Fall 2010).

ance of enrollment is at the pre-college level, with a persistent percentage of about 57 percent of all enrollments in mathematics units over the last two decades. Other survey results reveal only modest changes in the distribution of enrollment across the mathematics curriculum since 1990 (not shown in tabular form), with a small but persistent decline in enrollment in pre-calculus (college algebra and trigonometry) courses since 1995, a slight drop but also fluctuation in calculus enrollment from 1990 to 2010, and a modest increase in enrollment in statistics courses since 1990 and in other mathematics courses since 1995, including classes for non-mathematics majors (e.g., mathematics for liberal arts and mathematics for elementary school teachers).

The CBMS also examined faculty and instruction, which is an important issue for two-year colleges where part-time faculty are well documented and an important part of the teaching workforce (Townsend and Twombly, 2008). The Fall 2005 survey reveals the extent to which part-time faculty are engaged in mathematics instruction (Lutzer et al., 2007), with the percentage of two-year college mathematics sections taught by part-time faculty being 44 percent. (Although a percentage was not apparent in preliminary results from the Fall 2010 survey, it is clear that part-time faculty numbers remain high in the 2010 CBMS survey.) Part-time faculty members are most evident at the pre-college level, with 56 percent of these sections being taught by part-time faculty, and part-time faculty are least involved in teaching of mainstream calculus and advanced mathematics compared to other courses in the two-year mathematics curriculum, with only 12 percent and 9 percent, respectively.

These results suggest students taking the advanced college-level mathematics curriculum are most likely to be taught by a full-time faculty, a finding that seems to recognize the importance of advanced mathematics curriculum being taught by professionally trained mathematics specialists as well as the need to align standards with disciplinary requirements that support student progression (transfer) to the university. Noting this advantage, there is little evidence to suggest students who take even the most advanced two-year college mathematics courses intend to continue their study of mathematics at the university level as mathematics or STEM majors. Looking at all the CBMS data for 2005, Lutzer et al. (2007) suggested relatively few two-year college students intend to transfer to the university and major in mathematics at the four-year college level, although community college science and engineering (S&E) students do transfer to the baccalaureate level to pursue S&E and engineering technology baccalaureate degrees. Indeed, associates' degrees in S&E and engineering technology constitute about 11 percent of all associate's degrees awarded in 2007 (National Science Board, 2010), and presumably most of these students are interested in transferring. Looking retrospectively, about 44 percent of S&E graduates attended community colleges (Tsapogas, 2004). The proportion of associate's degree holders who are racial-ethnic minorities is higher among associate's degree than bachelor's degree holders in S&E fields, making these programs a rich ground for recruiting of STEM majors at the baccalaureate level (Handel, 2011). From 1995 to 2007, the number of S&E associate's degrees earned by racial-ethnic minority students more than doubled from 7,836 to 19,435.

The Fall 2005 CBMS survey data report on instructional approaches that provide insights into how two-year mathematics curriculum courses are taught (Lutzer et al., 2007). These data show over three-quarters of on-campus sections of college algebra and trigonometry, two courses core to the mathematics curriculum for many transfer students, are taught using the standard lecture method. The standard lecture method was less evident in pre-college course sections such as arithmetic (64%) and more evident in mainstream and nonmainstream calculus, elementary statistics, differential equations and technical mathematics (calculus), ranging from 81 percent to 93 percent. Given that calls for reform of mathematical pedagogy have been made for many years (see, for example, Wubbels and Girgus, 1997, and the authors of two-year college standards-based reform mentioned in the next section), it is perplexing that so little change has occurred in the teaching of such important two-year college mathematics courses.

Looking at both instructional and outreach methods in the Fall 2005 and Fall 2010 CBMS surveys (E. Kirkman and R. Blair, personal communication, December 4, 2011; Lutzer et al., 2007), results show college placement testing in mathematics is nearly universal in two-year colleges, although the preliminary finding from CBMS 2010 shows a 7 percent drop from 2005 to 2010 that deserves further study. Blair indicated she and her colleagues are still exploring the reason for this drop, but point out that all of the 90 percent of two-year colleges that report diagnostic testing report requiring mathematics placement testing for all incoming students. These results also show an increase in K-12 outreach opportunities and undergraduate research, but a decline in honors sections and special programs to encourage women and minorities to enroll in two-year college mathematics. In terms of the use of distance and online learning, the CBMS 2010 surveys show relatively modest use of online instruction, with most courses showing less than 10 percent of the sections using online learning systems. These results are consistent with previous results from CBMS 2000 and CBMS 2005.

Given the dearth of information about student enrollments and outcomes in two-year college mathematics, results of a national study by Horn and Li (2009) on postsecondary awards (credentials) below the baccalaureate level provide some insights into the scope and status of degree attainment. Most importantly, this research shows only .5 percent of all subbaccalaureate awards conferred in 2007 by Title IV postsecondary institutions (public community colleges as well as private for-profit institutions that award credentials below the baccalaureate level and also participate in federal financial aid) are in mathematics and science fields. Whereas this statistic is alarmingly low, it represents an 11.5 percent increase from 1997 to 2007. Also, in spite of the fact that females earn a majority of all subbaccalaureate credentials, enrollments in mathematics and science by males and females shifted toward males, from 46.9 percent in 1997 to 50.3 percent in 2007. Results on subbaccalaureate awards were not disaggregated by discipline and race; however, Horn and Li's study confirms that most subbaccalaureate awards are conferred to whites. Among minority groups, Hispanics showed the largest increase (75%) in overall subbaccalaureate-conferred awards from 1997 to 2007; followed by African Americans, with a 54 percent increase; and whites with an 11 percent increase only. Results of national studies by Tsapogas (2004) and Dowd, Malcom, and Bensimon (2009) also report the tendency of Hispanics to use the two-year college to pursue science and engineering degrees, though again, these results were not specific to mathematics.

Given the need to reach more women, minorities, and other underserved student populations, it is not surprising that professional mathematics groups have made recommendations to change mathematics education. Among various groups to respond to calls for reform, the American Mathematical Association of Two-Year Colleges (AMATYC) played an especially important leadership role, as this next section describes.

REFORM OF THE TWO-YEAR MATHEMATICS CURRICULUM

Following a wave of reform agendas at the K-12 level, mathematics professionals associated with two-year colleges eagerly jumped into the discussions. For many years, AMATYC has provided guidance to professionals who teach in two-year colleges throughout the United States. Consistent with efforts to improve mathematics education at the K-12 level, most notably the National Council on Teachers of Mathematics (2010) reform agenda on *Principles and Standards for School Mathematics*, AMAYTC's *Crossroads in Mathematics* (Cohen, 1995), and the subsequent *Beyond Crossroads* (Blair, 2006) initiatives conceived the two-year college position on reforming mathematics curriculum and instruction in community and technical colleges. The premise of "crossroads" identified in the *Crossroads in Mathematics* (Cohen, 1995) is that more citizens need to be prepared for STEM-oriented careers, including the "mathematics, science, engineering, and technology" workforce, but many students who seek a postsecondary education are not adequately prepared to perform at the college level. With respect to college preparation in mathematics, the *Crossroads in Mathematics* report anchored college readiness in students' needing the mathematics fundamentals as well as their needing to advance to studying calculus. A statement that captured the major focus of the *Crossroads in Mathematics* report is that:

> More students are entering the mathematics "pipeline" at a point below the level of calculus, but there has been no significant gain in the percentages of college students studying calculus (Albers et al., 1992). The purpose of *Crossroads in Mathematics* is to address the special circumstances of, establish standards for, and make recommendations about introductory college mathematics. The ultimate goals of this document are to improve mathematics education and encourage more students to study mathematics.

The AMATYC taskforce associated with *Crossroads in Mathematics* called for a "flexible framework for the complete rebuilding of introductory college mathematics" (p. 5), emphasizing student growth in knowledge of mathematics by enhancing the meaning and relevance of mathematics, the importance of laboratory teaching of mathematics, the use of technology as essential to up-to-date curriculum, "balanced" content and instructional strategies that include "viable components of traditional instruction," the contribution of mathematics to students' educational and career options, and the inclusion of diverse students (pp. 4-5).

The report also noted three related yet distinct sets of standards: standards for intellectual development, standards for content, and standards for pedagogy. Standards for intellectual development pertain to problem solving, modeling, reasoning, etc.; standards for content address such topics as real numbers and basic properties, solving linear equations, whole-number exponents, quadratic equations, etc.; and standards for pedagogy include teaching with technology, interactive and collaborative learning, connecting with other experiences, and experiencing mathematics (Cohen, 1995). The approach taken in the *Crossroads in Mathematics* report was that all students should grow in their fundamental knowledge of the normative mathematics content, supplemented with probability and statistics.

AMATYC's second standards document, entitled *Beyond Crossroads* (Blair, 2006), extended the goals, principles, and standards set forth in the earlier *Crossroads in Mathematics* report by calling for fuller and more strategic implementation of standards-based reform. *Beyond Crossroads* placed more emphasis on assessment of students' learning and promoting quantitative literacy (a topic discussed more fully later in this paper), meeting the needs of diverse learners, promoting active learning and online learning, promoting professionalism among full- and part-time instructors, and recognizing and involving more stakeholders in implementation of mathematics reforms. The *Beyond Crossroads* report acknowledged the complexity of reforming two-year college mathematics in ways not addressed in the first *Crossroads in Mathematics* text to facilitate "student learning and the learning environment" and encourage faculty, departments, and institutions to improve all facets of two-year college mathematics education" (p. 7). Whereas the normative mathematics course sequence was still dominant in the second *Beyond Crossroads* report, the content standards seemed to place more emphasis on the application of mathematics to solve problems and to collect, analyze, and use data to help faculty make informed decisions and grow as professionals. Blair (2006) also observed that the *Beyond Crossroads* standards were intended to enhance access to college for underserved students, noting that two-year college mathematics "holds the promise of opening paths to mathematical power and adventure for a segment of the student population whose opportunities might otherwise be limited" (p. 7).

College Renewal Across the First Two Years (CRAFTY) (Ganter and Barker, 2004) is a third initiative that has focused on improving two-year college mathematics, in this case by focusing on college algebra. CRAFTY recognizes that most college students enroll in college algebra to fulfill a general education requirement and never see the relevance of the subject to the rest of their college education. Very few of these students ever move beyond college algebra to enroll in calculus, a point consistent with the *Crossroads in Mathematics* report. CRAFTY is a subcommittee of the Committee on the Undergraduate Program in Mathematics (CUPM) at the Mathematical Association of America. The project looks at the introductory mathematics courses for the broad range of students who enroll in postsecondary education, most of whom will not be mathematics majors, and it solicits input from disciplinary groups (e.g., biology, engineering, computer science, etc.) on what mathematics departments can do to best prepare students for those disciplines. Ganter and Barker (2004) advanced the notion of a series of disciplinary-based workshops known as the Curriculum Foundations Project that were conducted across the country between 1999 and 2001 to discipline partners to state their views of the mathematics curriculum and engage with mathematics practitioners in a dialogue about ways to reform the curriculum. Through these and other interactive strategies, CRAFTY encouraged faculty to engage colleagues, college administrators, employers, and other local business leaders "to improve the role of College Algebra in our educational system and in the effectiveness of the present programs" (Small, 2002). Consistent with the CRAFTY approach, faculty are encouraged to use small group projects and technology applications that engage students in active use of mathematics to solve real-world problems.

A companion initiative to CRAFTY led by AMATYC, called "The Right

Stuff" (funded by the National Science Foundation), opened the dialogue within the two-year college mathematics community to re-envisioning college algebra and redesigning curriculum to meet the needs of students enrolled in college algebra who might not be calculus-bound. Through traveling workshops, AMATYC assists faculty to use materials that were directed at encouraging students to engage in "meaningful activities that promote the effective use of technology to support mathematics, further provide students with stronger problem-solving and critical thinking skills, and enhance numeracy" below the calculus level (American Mathematics Association for Two-Year Colleges, n.d.). AMATYC also administered an initiative called Mathematics Across the Community College Curriculum (MAC^3) (also funded by the National Science Foundation) that designed and shared materials that infused the mathematics curriculum with real-world problems and scenarios in collaboration with disciplines outside of mathematics (like science and economics) (R. Blair, personal communication, December, 8, 2011). Hillyard et al. (2010) summarized a number of studies from MAC^3 projects for a special theme of the AMATYC journal to demonstrate that mathematics can be connected to other disciplines, that students increase their quantitative literacy, that faculty members are positively impacted, and that "academic 'turf' conflicts that emerge when we move towards interdisciplinarity" (p. 7) are overcome.

Together, these several initiatives—the two AMATYC standards-based reforms and the CRAFTY, The Right Stuff, and the MAC^3 projects—represent important developments in improving two-year college mathematics curriculum in the United States. All of these initiatives have contributed to a national conversation to reform the curriculum. At a time when mathematics course enrollments have grown at an impressive rate in the two-year college, particularly during the decade of the 2000s, strong consensus has emerged about the need to improve two-year college mathematics as a means of enhancing the STEM pipeline in the United States. Given the importance of two-year college mathematics to the overall P-20 education agenda of the United States, it is important to examine two-year college mathematics innovations to lay a foundation for recommending the next steps for research, policy, and practice.

WHAT WE KNOW ABOUT INNOVATIVE APPROACHES TO TWO-YEAR MATHEMATICS

Due to the considerable enrollment in pre-college mathematics in two-year colleges, many innovations and reforms are focused on the pre-college (or developmental education) level. Studies of the struggles that students face enrolling in and navigating the sequence of pre-college mathematics courses have been immensely important to understanding student success (or lack thereof) in mathematics courses and the larger STEM pipeline. Researchers such as Bailey, Jeong, and Cho (2009); Perry et al. (2010); and many others have laid a foundation for understanding complex issues associated with pre-college mathematics. Despite a growing body of research, more rigorous research is needed on pre-college as well as college-level mathematics curriculum at the two-year college level. Research on mathematics teaching and learning at the classroom level is needed to provide a fuller and more nuanced understanding of content-based and pedagogically oriented reforms that may promote mathematics competency and positive student outcomes at the two-year level, including mastery of pre-college competencies, matriculating to and mastering advanced competencies, and advancing to and through the STEM pipeline.

An important recommendation that emerged from the AMATYC standards-based reform reports, particularly the 2006 *Beyond Crossroads* report, is consistent with the wider national and international conversations to emphasize quantitative literacy and quantitative reasoning as an element of or, in some cases, alternative to the normative mathematics course sequence. A leader in the dialogue about quantitative literacy, Steen (2001) argues that enabling students to use mathematics to solve real-world problems that are complex, ambiguous, and incomplete is the most important thing that college mathematics courses can do. She notes that "rarely will high school graduates be faced with problems that present themselves in the language of algebra" (Steen, 1992, n.p.), but just because students don't appreciate algebra in its traditional forms does not mean that it is not applicable or useful to them. Steen notes that quantitative literacy is rooted in real data that are part of life's diverse contexts and situations. She believes pedagogy should change to encourage quantitative thought that can help learners "to understand the meaning of numbers, to see the benefits (and risks) of thinking quantitatively about commonplace issues, and to approach complex problems with confidence in the value of careful reasoning" (Steen, 2001, p. 58). Students who experience quantitative literacy are empowered to think independently, to ask smart questions, and to confront complexities and challenges with confidence, and, as Steen concludes, "these are the skills required to thrive in the modern world" (p. 58).

Given the importance of this topic, it is unfortunate that the literature on quantitative literacy and quantitative reasoning is disconnected from literature on contextualized teaching and learning, integrated academic and technical curriculum, and problem-based learning. Referring to this collection of curricular and instructional approaches, Perin (2011) described contextualization as the "practice of systematically connecting basic skills instruction [in fields of study such as mathematics] to a specific content that is meaningful and useful to students" (p. 3). Her recent review of the literature includes findings of various types of contextualization employed in postsecondary settings, especially pre-college mathematics courses. Baker, Hope, and Karandjeff (2009) have explored the wide range of definitions that are used for contextualized instruction, and, to their credit, they link practices associated with contextualization to theories of learning and pedagogical strategies. Among the recommendations made by Baker et al. is the importance of exploring alternative formats for delivering the normative mathematics curriculum. Whereas rigorous research has not been performed on contextualized math at the two-year college level, an experimental study that examined the effect of training of high school math and career and technical education (CTE) teachers to work cooperatively to make math explicit in CTE classrooms produced statistically significant outcomes, including higher scores on standardized and college placement tests without negatively impacting technical skill attainment (Stone, Alfeld, and Pearson, 2008). This study has not been replicated in the two-year college context, but it would be very helpful to do so.

Examples of other innovative mathematics curriculum formats that are being studied include modularization, which involves delivering instruction in manageable segments or "chunks" (Rutschow and Schneider, 2011, p. 25), rather than traditional, semester-long courses. Mostly applied to pre-college mathematics, this strategy of chunking the curriculum could be extended to college-level mathematics. When implemented properly, students can achieve success in shorter time periods than traditional courses, which also motivates them to persist to the next shortened segment. Bailey et al. (2003) evaluated modularization in six NSF Advanced Technological Education (ATE) projects and four centers, and they noted that instructors praised the method for its flexibility and adaptability. The National Center for Academic Transformation Mathematics Emporium model, which Twigg (2011) described as a "silver bullet," combines modularization with technology-supported instruction (G. Reese and C. Kirby, personal communication, October 18, 2011).

Another innovation that is being attempted in mathematics, particularly pre-college mathematics, involves compression of the curriculum, meaning compressing the amount of time it takes for students to complete mathematics course sequences, and accelerating them toward their next course or completion. Compression often requires scheduling courses more hours a day for shorter amounts of time, and pairing courses that complement one another, including pairing mathematics and science courses or pairing multiple mathematics courses (including pre-college and college level) to create an intensive learning experience. Though most of the research on compression and acceleration is focused on the pre-college level curriculum, this strategy may be useful to attempt with college-level mathematics courses (e.g., college algebra and statistics). For example, two forms of acceleration were used by the FastStart Program at the Community College of Denver, wherein FastStart accelerated students through the mathematics course sequence by allowing students to enroll in a developmental course concurrently with a college-level course. Results from 11 student cohorts who began developmental mathematics at various levels revealed encouraging outcomes on retention and credit accumulation (Bragg, Baker, and Puryear, 2010). Synthesizing the literature on acceleration, Edgecombe (2011) noted evidence of the impact of acceleration on developmental education is limited but promising based on evaluations of FastStart and other similar programs.

Change of not only how mathematics is taught but also what is taught is also important for mathematics reformers. One of the most notable efforts in this regard are the Carnegie Foundation for Learning and the Dana Center's Quantway™ and Statway™ projects (Carnegie Foundation for the Advancement of Teaching, 2011a, 2001b) that are attempting to replace the normative pre-college mathematics courses with mathematics courses focused on quantitative literacy and statistics. Using an accelerated timeframe, the Quantway and Statway projects seek to prepare students for college level mathematics instruction. Quantway and Statway "enable developmental mathematics students in community colleges to complete a[n accelerated] credit-bearing, transferable mathematics course in one academic year while simultaneously building skills for long-term college success" (Cullinane and Treisman, 2010, p. 4). The Statway course sequence assists students to develop statistical literacy and engages them in mathematical reasoning using data, and it provides them with college credit in statistics. Cullinane and Treisman hypothesize that the adoption of a statistics sequence such as Statway will support many more students to engage in mathematical reasoning, especially when the curriculum is institutionalized from K-12 education and extended to the postsecondary level. Quantway is similarly focused on increasing the mathematical literacy of students who need to take pre-college mathematics by replacing traditional textbook-based, procedural instruction with numerical reasoning that is necessary to solve real-world problems. The Quantway pathway promotes an accelerated format, allowing students who place into elementary algebra to gain access to and move through a college-level quantitative reasoning course in one year.

Reform of instructional materials such as those associated with Statway and Quantway address a disconcerting problem that Kays (2004), Mesa (2010), and others have noted in the literature: the reliance on textbooks to structure and guide classroom instruction of mathematics.

These studies demonstrate the ways classroom teaching that relies on procedural-based textbooks reinforce the memorization of procedural knowledge at the expense of quantitative reasoning. Many mathematics texts are also tied to the normative math course sequence, and traditional pedagogies associated with teaching the mainstay courses in that sequence (e.g., algebra, geometry, trigonometry, and calculus) provide a valuable backdrop for a discussion of the critical needs that lay ahead as two-year college mathematics educators delve more deeply into reform. These studies suggest that systemic reforms, including curriculum, instruction, and instructional materials, are need to be pursued if mathematics education is to be responsive to the diversity of learners who seek the opportunity to pursue STEM-related programs of study in the two-year college.

RECOMMENDATIONS FOR FURTHER RESEARCH, POLICY, AND PRACTICE

In spite of strident pursuit of standards-based reform of two-year college mathematics, implementation of reform of the mathematics curriculum has been slow and uneven. National studies show more students are enrolling in two-year college mathematics, but a substantial portion of these enrollments are at the pre-college level, and many of these students never reach college-level mathematics. Thus, the STEM pipeline appears to be widening at the start, which is encouraging, but it also seems to narrow rapidly as students attempt to advance to college-level mathematics, a prerequisite to pursuing STEM programs of study and STEM-related careers.

To facilitate the role that two-year mathematics can play in providing access to the STEM pipeline and preparing larger numbers of postsecondary students, mathematics instruction needs to be sufficiently engaging and useful to support their interests and to assist them to make the commitment necessary to pursue a STEM program of study. A whole host of issues need to be addressed with respect to two-year college mathematics and the preparation of students who seek subbaccalaureate credentials and who desire to transfer to universities in STEM fields. Specific recommendations for research, policy, and practice to support this goal are described below.

A systemic, P-20 approach is needed to reform mathematics curriculum. Recommendations offered by a plethora of professional groups, including AMATYC and the Mathematics Association of American (MAA), and at different levels of the educational system are logical, reasonable, and substantive, and equally importantly, they consistently argue for a multi-level yet coordinated P-20 approach. Without such a strategic, collaborative endeavor, it will be difficult for two-year colleges that are caught between K-12 education and higher education to engage in reform, except in isolated ways. Given the national imperative to enhance the STEM pipeline, and the critical role that mathematics needs to play in that work, this recommendation may be the most important of all to emerge from the Summit on Realizing the Potential of Community Colleges as Pathways to STEM Education and Careers.

More research is needed to improve two-year college mathematics instruction. Although numerous pedagogical strategies are emerging that offer promise to change the way mathematics is taught at the two-year college level, CBMS survey data confirm the prevalence of lecture-led, teacher-centered instruction rather than the sorts of contextualized, problem- and project-based approaches that support quantitative literacy. Finding ways to support two-year college faculty to engage in professional development that reinforces innovative instructional reforms is important. Included in this list is the importance of helping faculty to adopt curriculum and instruction that draws upon students' everyday life experiences in the workforce, their communities, and other aspects of their lives. Mathematics instructors also need to understand how to integrate technologies to deliver instruction in the classroom or from a distance. Moreover, mathematics instructors need to understand how college placement tests can either impede or advance students through the mathematics curriculum. Involving faculty in decisions about assessment may help them to understand how college placement testing impacts student learning and ultimately, improves student outcomes.

More research is needed on the students who enroll in two-year college mathematics, especially college-level mathematics (college algebra and beyond), and how their experiences and performance in college-level mathematics courses influences subsequent enrollment, completion, and career decisions. Because two-year colleges are the gateway to postsecondary education for diverse learners, these schools have an important role to creating pathways that prepare students to advance to higher levels of postsecondary education. More research is needed to support the study of mathematics pathways, other than the normative mathematics sequence, and to understand how students "develop the 'habits of the mathematical mind' that are required to be successful in mathematics *and* science *and* engineering *and* technology courses" (R. Blair, personal communication, December 8, 2011). Students need to know what these new mathematical pathways look like and how they lead to STEM careers, and they cannot be expected to understand or navigate them on their own, without encouragement and support. Systemic change is needed to ensure that all students who have aspirations for STEM careers get the chance to learn mathematics in ways that fully and respectfully support their goals. If the nation expects more women and minorities to participate in STEM programs of study, fulfilling this recommendation is essential.

More and better data are needed to support practitioner engagement in active research on mathematics education at the local level, where two-year college mathematics faculty and other stakeholders engage in the teaching and learning process. Beyond participating in training, many two-year faculty would appreciate and benefit from opportunities to engage in active research that encourages them to try out new pedagogical strategies in the classroom and determine how they impact student learning. The Equity Scorecard™ *and* Benchmarking projects of the Center for Urban Education at the University of Southern California provide valuable examples of ways that professional development of two-year college faculty can be integrated with action research to address equity issues for minority students who seek to participate in STEM programs (Baldwin et al., 2011). The Achieving the Dream initiative has established a strong track record of engaging practitioners in using data to improve pre-college mathematics (Rutschow et al., 2011). Lessons learned from this initiative and other newer ones, such as Pathways to Results in Illinois (Bragg and Bennett, 2011), offer the potential to improve two-year college mathematics and support student success in STEM programs of study.

SUPPLEMENTAL INFORMATION

Methods

This paper relies on existing literature available from a number of sources. Most importantly, academic databases were queried to identify peer-refereed articles as well as books, monographs, reports, papers, and conference presentations on two-year college mathematics. Databases included in this review were ERIC, EBSCO, Education Full-Text, JSTOR, Dissertation Abstracts, and Sociological Abstracts. In addition, Google and Google Scholar were queried to identify relevant documents and materials that appear outside of the traditional scholarly databases. Searches of websites maintained by organizations known to research and publish on the topic of two-year college mathematics were conducted, including the National Center for Education Statistics, the National Science Foundation, the Community College Research Center at Teachers College, and Charles A. Dana Center at the University of Texas at Austin, the AMATYC website, and others. Keywords used in these searches included the following words singularly and in combination with one another: math, mathematics, mathematics education, developmental, remedial, pre-college, algebra, calculus, advanced mathematics, statistics, etc.

Keywords used to understand how the scholarly literature situates two-year mathematics curriculum in the broader liberal arts and sciences context included the following: liberal arts and sciences, liberal arts, science, STEM, STEM education, technology, technology education, engineering, engineering education, technician education, etc. Also, to ensure that the full spectrum of literature on two-year colleges was included in this literature review, I used an extensive set of keywords to capture the institutional context, including the following: two-year college, community college, technical college, and junior college. I also entered keywords related to four-year college and university, transfer, and articulation to determine whether literature was available to compare the two-year context to the four-year context, including transfer.

In addition to the above methods, I reached out to several two-year college mathematics experts, including David Lutzer, Ellen Kirkman, and Rikki Blair, all authors of the Fall 2005 and/or Fall 2010 CBMS surveys. Rikki Blair also served as editor of the 2006 *Beyond Crossroads* report of AMATYC and was an especially thoughtful and gracious contributor. I also sought guidance from several two-year college mathematics practitioners and colleagues at the University of Illinois, including George Reese, director of the Office of Mathematics, Science and Technology Education, and Catherine Kirby, assistant director of the Office of Community College Research and Leadership, who collaborated recently on a literature review on this same topic and brought numerous sources on two-year college mathematics to my attention. Finally, I offer my gratitude to Dr. Julia Makala, research specialist at the Office of Community College Research and Leadership, who offered a critical review that was invaluable to the final draft of this paper.

REFERENCES

Adelman, C. (2004). *Principal indicators of student academic histories in postsecondary education, 1972-2000*. Washington, DC: U.S. Department of Education, Institute of Education Sciences. Available: http://www2.ed.gov/rschstat/research/pubs/prinindicat/prinindicat.pdf [June 25, 2012].

Albers, D.J., Loftsgaarden, D., Rung, D., and Watkins, A. (1992). *Statistical abstracts of undergraduate programs in the mathematical sciences and computer science in the United States, 1990-1991 CBMS Survey*. (MAA Notes Number 23). Washington, DC: Mathematical Association of America.

American Mathematics Association for Two-Year Colleges. (n.d.). *The right stuff: Appropriate mathematics for all students*. Memphis, TN: Author. Available: http://www.therightstuff.amatyc.org/ [June 25, 2012].

Arbona, C., and Nora, A. (2009). The influence of academic and environmental factors on Hispanic college degree attainment. *The Review of Higher Education, 30*(3), 247-269.

Arnold, R. (2010). *Contextualization toolkit: A tool for helping low-skilled adults gain postsecondary certificates and degrees*. Boston, MA: Jobs for the Future. Available: http://www.jff.org/sites/default/files/BT_toolkit_June7.pdf [June 25, 2012].

Attewell, P., Lavin, D., Domina, T., and Levey, T. (2006). New evidence on college remediation. *The Journal of Higher Education*, 77(5), 887-924.

Bailey, T., Matsuzuka, Y., Jacobs, J., Morest, V.S., and Hughes, K. (2003). *Institutionalization and sustainability of the National Science Foundation's Advanced Technological Education Program*. New York: Community College Research Center, Teachers College, Columbia University.

Bailey, T., Jeong, D.W., and Cho, S. (2010). Referral, enrollment, and completion in developmental education sequences in community college. *Economics of Education Review*, 29(2), 255-270.

Baker, E., Hope, L., and Karandjeff, K. (2009). *Contextualized teaching and learning: A faculty primer*. Sacramento, CA: The Chancellor's Office of the California Community Colleges. Available: http://www.cccbsi.org/Websites/basicskills/Images/CTL.pdf [June 25, 2012].

Baldwin, J., and the Developmental Mathematics Committee. (1975). *Survey of developmental mathematics courses at colleges in the United States*. Garden City, NY: American Mathematical Association of Two-Year Colleges. Available: http://www.eric.ed.gov/PDFS/ED125688.pdf [June 25, 2012].

Baldwin, C., Bensimon, E.M., Dowd, A.C., and Kleiman, L. (2011). Measuring student success. *New Directions for Community Colleges*, 153(spring), 75-88.

Blair, R. (Ed.). (2006). *Beyond Crossroads: Implementing mathematics standards in the first two years of college*. Memphis, TN: American Mathematical Association of Two-Year Colleges. Available: http://www.amatyc.org/Crossroads/CRRV6/BC_V6_home.htm [June 25, 2012].

Bragg, D.D., and Bennett, S. (2011). *Introduction to pathways to results*. Champaign, IL: University of Illinois, Office of Community College Research and Leadership. Available: http://occrl.illinois.edu/files/Projects/ptr/Modules/PTR%20Intro%20Module.pdf [June 25, 2012].

Bragg, D.D., Baker, E.D., and Puryear, M. (2010). *2010 Follow-up of Community College Denver FastStart Program*. Champaign, IL: University of Illinois, Office of Community College Research and Leadership. Available: http://occrl.illinois.edu/files/Projects/breaking_through/FastStart_Final.pdf [June 25, 2012].

Carnegie Foundation for the Advancement of Teaching. (2011a). *Quantway*. Available: http://www.carnegiefoundation.org/quantway [June 25, 2012].

Carnegie Foundation for the Advancement of Teaching. (2011b). *Statway*. Available: http://www.carnegiefoundation.org/statway [June 25, 2012].

Cohen, A. (1984, July). *Mathematics in today's community college*. Paper presentation at the Sloan Foundation Conference on New Directions in Two-Year College Mathematics in Atherton, CA. ERIC Reproduction no. Ed 244 656. Available: http://www.eric.ed.gov/PDFS/ED244656.pdf [June 25, 2012].

Cohen, A., and Brawer, F. (1982). *The American community college*, 1st ed. San Francisco: Jossey-Bass.

Cohen, A., and Brawer, F. (1987). *The collegiate function of community colleges*. San Francisco: Jossey-Bass.

Cohen, A., and Ignash, J. (1992). Trends in the liberal arts curriculum. *Community College Review*, 20(2), 50-60.

Cohen, A., and Ignash, J. (1994). An overview of the total college curriculum. *New Directions for Community Colleges*, 86 (Summer), 13-29.

Cohen, D. (Ed.) (1995). *Crossroads in mathematics: Standards for introductory college mathematics before calculus*. Memphis, TN: American Mathematical Association of Two-Year Colleges. Available: http://beyondcrossroads.amatyc.org/doc/CH1.html [June 25, 2012].

Contemporary College Mathematics. (n.d.) Available: http://www.contemporarycollege algebra.org/index.html [June 25, 2012].

Cullinane, J., and Treisman, P.U. (2010). *Improving developmental mathematics education in community colleges: A prospectus and early progress report on the Statway Initiative.* Paper presentation at the National Center for Postsecondary Research (NCPR) Developmental Education Conference: What Policies and Practices Work for Students? Available: http://www.utdanacenter.org/downloads/spotlights/CullinaneTreismanStatway Paper.pdf [June 25, 2012].

Dowd, A.C., Malcom, L.E., and Bensimon, E.M. (2009, December). *Benchmarking the success of latina and latino students in STEM to achieve national graduation goals.* Los Angeles: University of Southern California, Center for Urban Education.

Edgecombe, N. (2011, May). *Accelerating the academic achievement of students referred to developmental education.* CCRC Brief No. 55. New York: Community College Research Center, Teachers College, Columbia University. Available: http://ccrc.tc.columbia.edu/Publication.asp?UID=920 [June 25, 2012].

Ganter, S.L., and Barker, W. (Eds.). (2004). *Curriculum foundations project: Voices of the partner disciplines.* Washington, DC: Mathematical Association of America. Available: http://www.maa.org/cupm/crafty/Chapt1.pdf [June 25, 2012].

Greene, J., and Forster, G. (2003). *Public high school graduation and college readiness rates in the United States.* New York: Manhattan Institute, Center for Civic Information. Available: http://www3.northern.edu/rc/pages/Reading_Clinic/highschool_graduation.pdf [June 25, 2012].

Grubb, W.N. (1999). *Honored but invisible: An inside look at teaching in community colleges.* New York: Routledge.

Handel, S. (2011, July). *Improving student transfer from community colleges to four-year institutions—The perspective of leaders from baccalaureate-granting institutions.* New York: The College Board. Available: http://advocacy.collegeboard.org/sites/default/files/11b3193transpartweb110712.pdf [June 25, 2012].

Hillyard, C., Korey, J., Leoni, D., and Hartzler, R. (2010, February). Math across the community college curriculum: A successful path to quantitative literacy. *MathAMATYC Educator, 1*(2), 4-9.

Horn, L., and Li, X. (2009, November). *Changes in postsecondary awards below the bachelor's degree: 1997 to 2007.* Washington, DC: National Center for Education Statistics, Institute of Education Sciences, U.S. Department of Education. Available: http://nces.ed.gov/pubs2010/2010167.pdf [June 25, 2012].

Kays, V. (2004). *National standards, foundation mathematics and Illinois community colleges: Textbooks and faculty as the keepers of content.* (Doctoral Dissertation). Office of Community College Research and Leadership, University of Illinois at Urbana–Champaign. Available: http://occrl.illinois.edu/publications/dissertation/2004/3 [June 25, 2012].

Koos, L. (1924). *The junior college.* Minneapolis, MN: University of Minnesota Press.

Lutzer, D., Rodi, S., Kirkman, E., and Maxwell, J. (2007). *Statistical abstract of undergraduate programs in mathematical science in the United States, Fall 2005 CBMS Survey.* Providence, RI: American Mathematical Society. Available: http://www.ams.org/profession/data/cbms-survey/full-report.pdf [June 25, 2012].

Medsker, L. (1960). *The junior college: Progress and prospect.* New York: McGraw-Hill.

Mesa, V. (2010). Examples in textbooks: Examining their potential for developing metacognitive knowledge. *MathAMATYC Educator, 2*(1), 50-55.

National Academy of Sciences, National Academy of Engineering, and Institute of Medicine. (2006). *Rising above the gathering storm: Energizing and employing America for a brighter economic future.* Committee on Science, Engineering, and Public Policy. Washington, DC: The National Academies Press.

National Council of Teachers of Mathematics. (2000). *Principles and standards for school mathematics.* Reston, VA: Author. Available: http://www.nctm.org/standards/ [June 25, 2012].

National Science Board. (2010). *Science and engineering indicators 2010*. Arlington, VA: National Science Foundation. Available: http://www.nsf.gov/statistics/seind10/pdf/front.pdf [June 25, 2012].

Perin, D. (2011). *Facilitating student learning through contextualization: A review of evidence*. Columbia University, Teachers College, Community College Research Center. Available: http://ccrc.tc.columbia.edu/Publication.asp?UID=954 [June 25, 2012].

Perry, M., Bahr, P.R., Rosin, M., and Woodward, K.M. (2010). *Course-taking patterns, policies, and practices in developmental education in the California Community Colleges*. Mountain View, CA: EdSource.

Provasnik, S., and Planty, M. (2008). *Community colleges, Special supplement to The Condition of Education, 2008*. (NCES 2008-033). Washington, DC: U.S. Department of Education, National Center for Education Statistics.

Rutschow, E.Z., and Schneider, E. (2011). *Unlocking the gate: What we know about improving developmental education*. New York: MDRC. Available: http://www.mdrc.org/staff_publications_386.html [June 25, 2012].

Rutschow, E. Z., Richburg-Hayes, L., Brock, T., Orr, G. Cerna, O., Cullinan, D., Kerrigan, M. R., Jenkins, D., Gooden, S., and Martin, K. (2011, February). *Turning the tide: Five years of Achieving the Dream in community colleges*. New York: MDRC. Available: http://www.mdrc.org/publications/578/overview.html [June 25, 2012].

Small, D. (2002). *An urgent call to improve traditional college algebra programs*. Available: http://www.contemporarycollegealgebra.org/national_movement/an_urgent_call.html [June 25, 2012].

Steen, L.A. (1992). Does everybody need to study algebra. *Mathematics Teacher, 85*(4), 258-260. Available: http://www.stolaf.edu/people/steen/Papers/everybody.html [June 25, 2012].

Steen, L.A. (2001). Quantitative literacy. *Education Week on the Web, 21*(1), 58.

Stone, J., Alfeld, C., and Pearson, D. (2008). Rigor and relevance: Enhancing high school students' math skills through career and technical education. *American Educational Research Journal, 45*(3), 767-795.

Townsend, B., and Twombly, S. (2008). Community college faculty: What we know and need to know. *Community College Review, 36*, 5-24.

Tsapogas, J. (2004, April). *The role of community colleges in the education of recent science and engineering graduates*. InfoBrief (NSF 04-315). Arlington, VA: National Science Foundation.

Twigg, C.A. (2011, May/June). The math emporium: Higher education's silver bullet. *Change Magazine (online)*. Available: http://www.changemag.org/Archives/Back%20Issues/2011/May-June%202011/math-emporium-full.html [June 25, 2012].

U.S. Census Bureau. (2012). *The 2012 statistical abstract: The national data book*. Washington, DC: Author.

Waits, T., Setzer, J.C., and Lewis, L. (2005). *Dual credit and exam-based courses in U.S. public high schools, 2002-03*. Washington, DC: U.S. Department of Education, National Center for Education Statistics. Available: http://nces.ed.gov/pubs2005/2005009.pdf [June 25, 2012].

Wubbels, G., and Girgus, J. (1997). The natural sciences and mathematics. In J. Gaff and R. Ratcliff and Associates (Eds.), *Handbook of the undergraduate curriculum* (pp. 280-300). San Francisco: Jossey-Bass.

Appendix D

Developing Supportive STEM Community College to Four-Year College and University Transfer Ecosystems

Alicia C. Dowd*
Associate Professor, Rossier School of Education,
and Co-Director, Center for Urban Education
University of Southern California

EXECUTIVE SUMMARY

Two-year to four-year college and university transfer pathways in science, technology, engineering, and mathematics (STEM) fields are too narrow and must be expanded to meet the social and economic demand in the United States for a greater number and a more diverse membership of scientists, engineers, and technicians. Faculty members have a critical role to play in expanding STEM transfer pathways. The value of structural, informational, and policy solutions, such as state and institutional articulation agreements, transfer information websites, state longitudinal data bases, and the accountability reporting made possible by such data, should be strengthened through initiatives to change the "culture of science" in ways that will foster culturally inclusive pedagogy and practices.

Any form of cultural and deep-seated organizational change requires a concerted effort over an extended period of time. Such an effort requires thought leaders, strategic communications, dedicated "change agents," and a growing perception that norms are changing for the good. Prominent STEM scholars and educational leaders have recently provided a blueprint for change in comprehensive national reports, including the National Science Board's *Preparing the Next Generation of STEM Innovators: Identifying and Developing our National Human Capital* and the National

---

*With valuable research assistance provided by Svetlana Levonisova, Raquel Rall, Cecilia Santiago, Misty Sawatzky, and Linda Shieh.

Academies' *Expanding Underrepresented Minority Participation: America's Science and Technology Talent at the Crossroads.*

The recommendations of these reports emphasize the need for greater access for all students to academic excellence in STEM and the necessity of improving talent assessment systems in order to identify currently overlooked abilities. Transfer admissions in general and in STEM in particular are particularly hampered by poor signaling of student talents and accomplishments because the quality of the community college curriculum is viewed with suspicion by university and liberal arts faculty. To address this problem, the National Science Board's recommendation to foster a supportive ecosystem is paramount. Creating a supportive ecosystem for transfer students requires the formulation of new incentives and rewards for college faculty in all sectors as well as professional development in teaching, curriculum development, and collaboration. Such professional development activities will be well received if they are accorded prestige and allocated time and resources for the production of new knowledge through research, design experiments, and inquiry, which is the systematic use of data, reflection, and experimentation to improve professional practices.

Taking into account the prestige associated with success in STEM fields and the generally separate nature of faculty networks in different sectors and disciplines, this report endorses the following recommendations:

(i) Create Evidence-Based Innovation Consortia (EBICs), involving STEM faculty, deans, and department heads in geographic and market-based groupings of two-year and four-year colleges and universities to review, invent, experiment with, and evaluate innovative curricula, pedagogies, and assessments of student talents and learning.

(ii) Devote institutional, private, and federal funds to STEM-specific work-study awards and transfer scholarships for transfer students and charge EBICs with the recruitment and selection process.

(iii) Develop a pool of eligible cohorts of students at community colleges through jointly administered two-year and four-year college learning communities and bridge programs, recruiting and retaining a diverse group of students using holistic admissions and assessment criteria developed through the EBICs.

(iv) Accord prestige to EBIC membership and the recipients of the transfer work-study awards and scholarships through high-profile communications and selection procedures.

CREATING MORE ROBUST STEM TRANSFER
PATHWAYS: NATIONAL CONTEXT

No single data source provides a comprehensive estimate, but the available evidence suggests two-year to four-year college and university transfer in STEM fields is small relative to the need for a greater number of STEM-educated citizens, workers, and professionals in the United States. The barriers and potential solutions to increasing access through transfer to STEM bachelor's and graduate degrees for transfer students are the subject of this report. This consideration takes place in a broader national context. In May 2010, as mentioned above, the National Science Board (NSB) issued its comprehensive report entitled *Preparing the Next Generation of STEM Innovators: Identifying and Developing Our National Human Capital*, and in 2011, the National Academies issued *Expanding Underrepresented Minority Participation: America's Science and Technology Talent at the Crossroads*. The three keystone recommendations of the *Next Generation* report (National Science Board, 2010) and several of its policy actions deserve particular attention when examining the evolving relationships between community colleges and four-year colleges and universities for the purpose of broadening STEM transfer pathways. These are

(1) NSB Keystone Recommendation #1: Provide opportunities for excellence
(2) NSB Keystone Recommendation #2: Cast a wide net
   (a) Policy Action: Improve talent assessment systems
   (b) Policy Action: Improve identification of overlooked abilities
(3) NSB Keystone Recommendation #3: Foster a supportive ecosystem
   (a) Policy Action: Professional development for educators in STEM pedagogy

These particular recommendations and policy actions, excerpted from among others in the NSB's *Next Generation* (2010) report, are highlighted here because the challenges of (1) providing quality science and mathematics teaching to all students (i.e., "opportunities for excellence"), (2) improving assessment and talent identification, and (3) creating supportive ecosystems through professional development for STEM educators are particularly central to the challenge of creating more robust STEM transfer pathways. They are also essential in light of the urgency articulated in the *Crossroads* report (National Academy of Sciences, National Academy of Engineering, and Institute of Medicine, 2011) to substantially increase the racial-ethnic diversity of participation in STEM fields. The dimensions of these problems are cultural as well as structural; yet prevailing attempts to improve transfer, such as articulation agreements, curriculum alignment through common course numbering, and policies guaranteeing transfer of credits, have most often been structural. However, to improve transfer in STEM, it will be necessary to consider the cultural characteristics of STEM learning environments and those who have traditionally succeeded in them in formal educational systems in the United States.

Before discussing the culture of science and how it pertains to the issue of the improvement of transfer access to STEM bachelor's and graduate degrees (see section III below), I first present statistics to provide a sense of the potential supply of STEM transfers and sources of data to estimate the number of transfers in STEM fields (section I). Then, I briefly review the barriers and potential solutions to improve transfer access from community colleges (section II). The report then concludes with discussion of recommendations to create Evidence-Based Innovation Consortia (EBICs) as place- and market-based entities with a focus on improving STEM transfer pathways (section IV).

I. POPULATION AND TRENDS IN THE NUMBER OF POTENTIAL STEM TRANSFER STUDENTS

Data provided by the National Center for Education Statistics (NCES) include the total number of credential-seeking undergraduates, distinguishing those enrolled in subbaccalaureate programs from those enrolled in bachelor's degree programs. In 2007-2008, the subbaccalaureate population numbered 9,822,000, with 6,383,000 classified as enrolled in career education, 2,361,000 enrolled in academic education, and the remainder undeclared (National Center for Education Statistics, n.d.-b). Career education includes some technical fields such as agricultural and natural resources, computer and information services, engineering, and health services, as well as non-STEM fields such as business management, communication and design, and legal and social services. Vocational degrees such as cosmology and protective services are also included. Academic education includes general education courses in science and mathematics. These numbers represent students in public two-year colleges (community colleges) and in for-profit, proprietary colleges combined. In the very broadest terms, these nearly 10 million students represent the total potential pool of transfer students. In Fall 2008, the count of students enrolled in community colleges for credit numbered 7.4 million (Mullin, 2011).

However, many of these students are strictly seeking vocational training, do not aspire to transfer, and earn certificates in short-term programs rather than associate's degrees (Mullin, 2011). The growing interest in applied baccalaureate degrees (Ruud and Bragg, 2011) notwithstanding, the nearly two-to-one ratio of students in career education versus academic education reflected in the figures above indicates that the majority of students enrolled at the subbaccalaureate level are earning credits in vocational courses that would not count toward a bachelor's degree.

   The American Association of Community Colleges reports on degrees awarded by public two-year institutions. In 2009-2010, approximately one million degrees and certificates were awarded, including 630,000 associate's degrees (Mullin, 2011, p. 6). Of these, 40 percent were classified as degrees in the liberal arts and sciences or humanities, which align with a general education focus within a transfer-directed curriculum. The number of associate's degrees awarded by community colleges represents an overall increase of 86 percent from two decades earlier, but growth rates were much higher for Hispanics (383%), blacks (204%), and Asian-Pacific Islanders (APIs, 230%) (Mullin, 2011, pp. 17-18). Currently, Hispanics, American Indians and Alaska Natives, and African Americans all earn associate's degrees at higher rates than white and Asian-Pacific Islander students. For example, in 2007-2008, 36 percent of degrees earned by Hispanics and 30 percent earned by blacks were associate's degrees, compared to 23 percent for whites and 19 percent for APIs. Conversely, bachelor's degree completion rates were lower, with only 11 percent of Hispanics in the 25-29 year age group having at least a bachelor's in 2008 and 17 percent of blacks. These figures compare with 33 percent of whites and 60 percent of APIs in the same age group (Aud, Fox, and KewalRamani, 2010). NCES (2011) reports that 14.4 percent of all students who began their studies in public two-year institutions earned an associate's degree within the six-year period of 2004-2009.

   Certificates were awarded at community colleges for programs ranging from less than one year to four years in duration. The increase in certificates was much greater than the growth in associate's degrees, growing 776 percent and 338 percent for Hispanic and black students, respectively (Mullin, 2011, pp. 17-18). This trend mirrors the increases in degrees and certificates awarded by for-profit postsecondary institutions, which has been the fastest growing sector of higher education over the past decade, with enrollments doubling from 192,000 to 385,000 from 2000 to 2009 (National Center for Education Statistics, 2011). These numbers are significant because they show a shift in demand for subbaccalaureate education away from community colleges toward the for-profit sector. Some attribute the rise of the for-profit sector to the inability of public colleges to meet the demand for higher education (Lee and Ranson, 2011). A notable part of the changing STEM education landscape is the growing number of students earning associate's degrees at for-profit institutions and the growth in the number of short-term certificates awarded in both sectors. Data from the NCES indicate that nationally the most popular STEM-related career education fields of study at the associate's degree level in 2007-2008 were health sciences, enrolling 1,627,000 students (and 21% of the total); engineering and architecture, enrolling 396,000 (6.7%); computer and information services, enrolling 336,000 (3.8%); and agriculture and natural resources, enrolling 50,000 (.7%) (National Center for Education Statistics, n.d.-c). The number of associate's degrees awarded in the health sciences in 2008-2009 represents a 77 percent increase over 1998-1999. Computer and information sciences also saw overall growth of nearly 34 percent during that time period, but nevertheless experienced a loss of 27 percent in the number of degrees awarded to women. Engineering and engineering technologies experienced a decline in degrees awarded of nearly 8 percent for men and women combined, but of 24 percent for women (National Center for Education Statistics, n.d.-a). Agriculture and natural resource fields experienced a decline among both men and women, with a nearly 14 percent loss overall. These trends mirror declining proportions of women in engineering and computer sciences at the bachelor's degree level (National Science Foundation, 2011).

Hardy and Katsinas (2010) investigated a longer period of time by analyzing institutional data captured by the annual snapshot of higher education in the Integrated Postsecondary Education Data System (IPEDS). They compared the number of associate's degrees awarded over three decades (1985–1986, 1995–1996, and 2005–2006), focusing on broad STEM codes including engineering, engineering technologies/technicians, biological and biological sciences, mathematics and statistics, physical sciences, and science technologies/technicians. The article focuses on gender, in particular, and shows that although the overall number of associate's degrees awarded in STEM is increasing, the percentage awarded to women is not.

Contested but Inadequate Transfer Rates

The estimation of transfer rates is contested (Horn and Lew, n.d.). Depending on how broad or restrictive the denominator is, the determination of who "counts" in estimating the rate and the length of time allowed for transfer to take place, transfer rates vary widely. A broad-based national estimate of the proportion of community college students who transfer to a four-year institution is 25 percent (Melguizo and Dowd, 2009). However, this number varies by state, socioeconomic status (SES), and students' demographic characteristics. Students from higher SES households are more likely to transfer than those from lower SES households, with a difference of 45 percentage points between the 10 percent transfer rate for low-SES students and the high end at 55 percent (Dougherty and Kienzl, 2006). Using a broad denominator of Latinos entering community colleges in California, Ornelas and Solorzano (2004) report an analysis of California Postsecondary Education Commission (CPEC) data indicating that only 3.4 percent of Latinos transfer to a California four-year public institution.

Another point of contention is whether transfer students experience a penalty in their pursuit of a bachelor's degree from starting at a community college. Utilizing statistical models to compare students of equivalent characteristics and qualifications, some find that there is a "diversion effect" (e.g., Cabrera, Burkum, and La Nasa, in press), by which transfer students become diverted from bachelor's degree attainment. Others find a "democratization effect," meaning that the open access community college ultimately democratizes access by providing an effective pathway to the bachelor's degree (Melguizo and Dowd, 2009).

Arbona and Nora (2007), analyzing National Longitudinal Educational Survey data of a sample initially collected in 1988 (NELS: 88), found that among those Latino students who first attended a community college, only 7 percent had obtained at least a bachelor's degree by 2000. Similarly, an estimate obtained from the Beginning Postsecondary Students Longitudinal Study (BPS:96/01) showed that although 25 percent of Hispanic students who attended a two-year college initially intended to transfer to a four-year institution and obtain a bachelor's degree, six years after first enrolling in community colleges only 6 percent had been awarded a bachelor's degree (Hoachlander, Sidora, and Horn, 2003). Notwithstanding these debates, few analyses conclude that transfer rates are high enough to fulfill the potential of community colleges to provide first generation, low-income, and underrepresented racial-ethnic minority group students with a satisfactory chance of earning a bachelor's degree.

An Initial Profile: Latina and Latino STEM Bachelor's Degree Holders Who Transferred

None of the studies and reports above provides estimates of the numbers of community college transfer students in STEM fields, revealing that further research is needed to produce such estimates. In this subsection, I present a brief profile of Latina and Latino STEM transfers based on a study conducted by the Center for Urban Education at USC with funding from the National Science Foundation to begin to fill this research gap. Transfer is of particular importance for increasing Latina and Latino participation in STEM because Latinas and Latinos are disproportionately enrolled in community colleges (Adelman, 2005), particularly in populous states with growing Latino populations, such as California, Florida, and Texas. Estimates vary, but roughly 60 percent of Latino students enrolled in postsecondary education attend a community college (Arbona and Nora, 2007; Snyder, Tan, and Hoffman, 2006).

Expanded transfer access is necessary because although Hispanic participation in STEM fields has risen, it has not kept pace with Hispanic population growth. Growth in the number of bachelor's degrees awarded to Hispanic students has occurred primarily in nonscience and engineering fields. From 1998 to 2007, there was a 64 percent increase in the number of nonscience and engineering bachelor's degrees awarded to Hispanic students, as compared to an increase of only 50 percent in science and engineering degrees awarded to Hispanic students. Further, the proportion of STEM doctoral degrees awarded to Hispanic students (estimated at less than 5 percent) severely lags the proportion of Hispanics in the U.S. population (around 15%).

Analyses conducted by Lindsey Malcom (2008a) and Alicia Dowd (Dowd, Malcom, and Macias, 2010) of the NSF's National Survey of Recent College Graduates (NSRCG:2003) present a portrait of the fields of study of Latina and Latino STEM[1] bachelor's degree holders who transferred from community colleges with associate's degrees, based on a sample of students who earned bachelor's degree in 2003. The analyses examine the fields of study in which Latino STEM bachelor's degree holders earned their degrees, comparing degrees awarded at Hispanic-Serving Institutions (HSIs) and those at non-HSIs.

Degrees awarded at HSIs (which are defined by enrollment of Hispanic students equal to or exceeding 25% of full-time students) and non-HSIs were differentiated because only 10 percent of institutions in the United States enroll the majority (54%) of Latino undergraduates (Horn, 2006). HSIs tend to be less selective nonresearch colleges and universities. Traditionally they have received less federal funding than research universities and selective institutions. Although nearly 40 percent of bachelor's degrees awarded to Latinas and Latinos in all fields of study are granted by HSIs (Santiago, 2006), that figure shrinks to 20 percent when the analysis is limited to STEM degrees (Malcom, 2008a; Malcom, Dowd, and Yu, 2010). This indicates that HSIs do not do as well at retaining Latinos in STEM fields as in other fields.

Our analysis of the NSCRG data, in which transfer students were defined as those who had first earned an associate's degree, showed that most transfer students who ultimately earn bachelor's degrees in STEM fields major in the social and behavioral sciences. This is true at HSIs, where these majors account for 60 percent of STEM baccalaureates, as well as at non-HSIs, where the share is 70 percent. There is one critical area of study in which HSIs graduate a substantially larger percentage of STEM transfers than non-HSIs. Of Latino STEM baccalaureates who graduate from HSIs, 18 percent earn their degrees in computer science and mathematics compared with only 5 percent of STEM transfer graduates

---

[1]The definition of STEM fields employed by the National Science Foundation includes computer science, mathematics, life sciences, physical sciences, behavioral and social sciences, and health-related fields.

at non-HSIs. On the other hand, HSIs appear to be lagging behind non-HSIs in terms of awarding bachelor's degrees to Latinos in the biological, agricultural, and environmental sciences (3% as opposed to 11%) and in engineering (1% as opposed to 7%).

These statistics present a portrait of Latino STEM transfer in which we see that (1) transfer pathways from community colleges are narrow; (2) the majority of degree holders who earned an associate's degree before earning a bachelor's degree in STEM earned their degrees in social and behavioral sciences, rather than in computer science, mathematics, biological, agricultural, and environmental sciences, engineering, physical science, or in fields designated as science and engineering related; (3) Latino students had a better chance of earning a STEM degree outside of the social and behavioral sciences if they did not earn an associate's degree first. These figures would change if we used a different definition of transfer students (for example, those who transferred after the equivalent of one year of study, or 30 credits), but they illustrate that certain pathways to STEM bachelor's degrees are not as readily accessible for students who start out in community colleges.

Clearly, similar portraits must be created for other groups of students. However, given that HSIs are typically nonselective four-year institutions and that Latino students are the fastest growing demographic group, this portrait of Latino transfer in STEM provides a good starting point for gaining an understanding that STEM transfer pathways are not nearly as robust as they need to be. Latino community college transfers who first earn associate's degrees have lower access to STEM bachelor's degrees at academically selective and private universities than their counterparts who do not earn an associate's degree prior to the bachelor's. Available studies of transfer trends, in which the analyses were not restricted to STEM fields or to Latinos, suggest that transfer has become more limited to selective institutions while fluctuating and leveling off in nonselective institutions during the 1980s and 1990s (Dowd, 2010; Dowd and Melguizo, 2008; Dowd et al., 2006). These results are not based on the most current data, but the forces that likely diminished transfer during those decades are still active today, including intensive demand for elite education that make transfer applicants less attractive to selective institutions (Dowd, Cheslock, and Melguizo, 2008). The loss of transfer access to selective institutions is of concern in regard to STEM graduate degree production because the competitive, "top 100" STEM research universities are the main gateways to STEM doctoral and professional degrees. As long as selective institutions restrict transfer access, the challenge of creating more robust transfer pathways in STEM for community college students will fall largely to nonselective institutions.

Generating Portraits of Transfer in STEM for Other Groups

How would the figures presented above change if the focal group of interest changed from Latina and Latino students to white and Asian students or to African Americans, Native Americans, women, students with disabilities, or other underrepresented groups? Replicating the results presented above for other groups of interest using the NSRCG data would be one way to answer this question. Arbona and Nora (2007) have analyzed the NELS database to examine transfer of Latino students; other researchers might conduct similar analyses, although they might encounter difficulties in estimation due to small sample sizes.

Wang (2011) has valuably proposed to examine transfer pathways in the new Educational Longitudinal Study (ELS) data, which will provide more current estimates disaggregated by a variety of demographic groups of interest. The ELS monitors a nationally representative cohort of students in their sophomore year of high school. In 2006, data about this sample were collected regarding the colleges the students applied to, the financial aid they received, and their postsecondary enrollment, among other information. In 2012, members of the cohort will be interviewed again to learn about their outcomes, including persistence and experience in higher education, and/or transitions into the labor market. Another, more specialized dataset may be especially useful for examining student pathways to and within engineering. The MIDFIELD database is a longitudinal database containing information from 11 public institutions for 226,221 students that have ever declared engineering as a major from 1988 through 2009. It includes data regarding student behaviors, including the majors they change to, and the major students subsequently graduate in. It contains student demographic information, history of courses taken, and grades received, as well as degrees awarded. Consequently, this dataset is a resource for mapping the types of paths students take after matriculating in engineering. It holds potential use for studying choices taken by students leaving engineering, and whether this group disproportionately comprises members of underrepresented student populations. In addition to information on first-time students admitted to college engineering programs, the MIDFIELD database also includes information regarding the pathways of transfer students who are admitted to engineering programs.

II. STRUCTURAL BARRIERS TO STEM TRANSFER AND PROMINENT SOLUTION STRATEGIES

Before moving into a discussion of cultural barriers to STEM transfer, it is important to acknowledge structural barriers to transfer and take stock of the most prominent contemporary strategies to broaden transfer pathways. The primary curricular barriers are lack of articulation of coursework in the two-year and four-year sectors; lengthy remedial, basic skills course sequences (particularly in mathematics); and the separation of special programs from the core curriculum. Challenges students encounter in financial aid and advising include "sticker shock" when contemplating four-year college and university prices, lack of information about the multiple sources of financial aid, poor access to counselors, and the lack of participation of faculty members in transfer advising.

Transfer and Articulation Policies Are Insufficient to Improve STEM Transfer Access

The goal of establishing curriculum "articulation" and alignment between the community college and four-year college and university curricula has been a policy focus for several decades. Although Zinser and Hanssen (2006), based on an analysis of national data from the Advanced Technological Education (ATE) program, conclude that articulation agreements for the transfer of two-year technical degrees to baccalaureate degrees are valuable, other analyses of secondary databases indicate that state-level articulation agreements have statistically insignificant effects on the likelihood that community college students will transfer (Anderson, Alfonso, and Sun, 2006; Anderson, Sun, and Alfonso, 2006; Kienzl, Wesaw, and Kumar, 2011).

These results indicate that articulation agreements are not likely to be effective on their own in substantially broadening STEM transfer pathways. California's recent experience in the early stages of implementing a guaranteed transfer degree, legislated in Fall 2011, illustrates some of the challenges in state policies intended to improve curriculum alignment. The new law stipulates that community colleges offer associate's degrees for transfer that the California State University (CSU) campuses would be obliged to accept. The mandated degree is 60 credits, including 18 credits in an area of academic focus that should provide a transfer student access to a similar major field of study at the university. The adoption of this law led to a process of negotiations between community college and university curriculum committees to identify articulated degree programs. By December of 2011, 16 associate's degrees were approved for transfer and priority admissions, but only two of these were in STEM fields (mathematics and physics) and about a third of the CSU campuses had yet to confirm availability of a matching degree program in those fields.

States have had varying success in using postsecondary policy to improve transfer pathways in STEM. Malcom (2008a, 2008b) illustrated this through analysis of the share of Latina and Latino STEM baccalaureates in NSF's 2003 National Survey of Recent College Graduates (NSRCG) who earned associate degrees. Examining the five states with the largest populations of Latinos—California, Florida, Illinois, New York, and Texas—she found that nearly half of all Latinos in Florida who were awarded a STEM bachelor's degree had also earned an associate's degree. This represents a much greater reliance on community colleges for STEM degree production in Florida, which has strong statewide articulation policies, than elsewhere. In New York, California, and Illinois, the share of Latino STEM bachelor's degree holders who had first earned associate's degrees (27.9%, 22.2%, and 16.3%, respectively) was closer to the national average of 20 percent. The proportion in states other than these five, which would include states with smaller community college systems, was considerably lower, at 9.2 percent.

The Substantial Challenge of Developmental Education in Mathematics

There is a growing recognition of the need to improve the teaching of foundational mathematics to young adults and adults in order to improve the persistence, degree completion, and transfer of community college students (Attewell et al., 2006; Bailey and Morest, 2006; Dowd, 2008; Grubb et al., 2011; Kirst, 2007; Levin and Calcagno, 2008). Many community college students are placed in classes, typically in mathematics, English, or writing, that do not carry credit towards an associate degree or bachelor's degree. These courses are referred to as remedial, basic skills, or developmental. Nationally, 42 percent of students enrolled in public two-year institutions in 2007–2008 took at least one remedial course, a share that is higher than in any other postsecondary sector. Eight percent of students required two remedial courses and 5 percent required three (Aud, Fox, and KewalRamani, 2011). Mathematics is the most common subject in which students require remediation, with national estimates hovering around 50 percent (Bahr, 2010; Parsad, Lewis, and Greene, 2003). Remedial testing and the long basic skills curriculum have disparate impacts on African Americans and Latinos, who are more likely to be placed in remedial courses and less likely to complete them successfully (Aud, Fox, and KewalRamani, 2011). By some national estimates, approximately half of black and Hispanic community college students earn remedial credits in mathematics (Bahr, 2010). In California, where the sheer size of the community college sector, with its 110 colleges, drives attention to community college issues, some estimate that 80 to 90 percent of students require remediation, with math being the greatest area of need (Grubb et al., 2011).

These statistics indicate that the challenge of remedial education is not unique to community colleges. However, the need for remediation and the often lengthy, skills-based remedial curriculum impedes students' ability to transfer or even contemplate transfer, as the lengthy time frame is discouraging. Students who test and are placed in courses such as arithmetic and pre-algebra, as well as remedial writing or English language courses, can face several semesters, or even years, of coursework that does not count for transfer to a four-year institution. Two recent studies from California are informative to illustrate the magnitude of the demand and the racial-ethnic equity implications of remedial education in community colleges. Hagedorn and colleagues (Hagedorn and DuBray, 2010; Melguizo, Hagedorn, and Cypers, 2008), analyzed transcripts of more than 5,000 students enrolled in the Los Angeles Community College District (LACCD) and examined basic skills mathematics course placements and completion. Over a third of students who had declared a STEM focus for their studies were initially placed in the lowest level course. Seventy-five percent of students were able to pass their first course on their first attempt. However, African American students were less likely to pass on the first attempt and both African American and Hispanic students emerged with lower mathematics GPAs. Mathematics appeared to pose a particular challenge; for example, African American students had equal rates of success to other student groups in science courses.

These results from Los Angeles are mirrored in California as a whole. Analyzing data from the California Community College Chancellor's Office for the Fall 2005 cohort, Bahr (2010) found that black and Hispanic students were disproportionately enrolled in mathematics basic skills courses and experienced low rates of successful remediation. His findings indicate that the rates of successful remediation in mathematics ranged from one-quarter to one-third of white and Asian students, in comparison to one-fifth of Hispanic and one-ninth of black students (Bahr, 2010, p. 232).

The equity implications of the remedial education challenge are evident given that lower income and underserved racial-ethnic minority students are less likely to receive adequate mathematics preparation in high school (Attewell et al., 2006; Bahr, 2010; Dowd, 2008) and less able to bear the opportunity costs of time spent in remediation (Melguizo, Hagedorn, and Cypers, 2008). Further, more affluent students can avoid strict remedial policies in the public sector by enrolling in private colleges and universities, where they receive stronger academic support to progress from remedial to degree-credit coursework. In addition, the reliability and validity of the placement tests have been questioned (Attewell et al., 2006; Brown and Niemi, 2007; Hughes and Scott-Clayton, 2011), in part because students with similar levels of academic preparation and test results can experience very different course placements, depending on their state of residence and their choice of institution within the same state.

Finally, it is not clear that students who are placed in remedial courses achieve better academic outcomes, in terms of persistence and credit accumulation, when compared with similarly qualified students who were not placed in developmental courses, because the available quasi-experimental evidence is mixed (Hughes and Scott-Clayton, 2011). All the available evidence indicates that the demand for mathematics basic skills education is substantial and that the current curriculum and instructional methods are not up to the task. The goal of improving transfer access to STEM degrees, therefore, is intertwined with the need to improve basic skills mathematics education.

Curricular and Programmatic Barriers and Potential Reforms

Educational researchers have conducted numerous case studies of transfer involving particular groups of students or institutions (Bensimon and Dowd, 2009; Bensimon et al., 2007; Cejda, 1998, 2000; Gabbard et al., 2006; Laanan, 1996; Lester, 2010; Ornelas and Solorzano, 2004; Townsend and Wilson, 2006). Common themes in the findings identify institutional barriers to transfer such as lack of information, confusing transfer curriculum requirements, the demands of remedial education (as noted above), and the struggle many students face becoming acclimated to a new campus environment at the four-year institution. Many students who transfer experience a "border crossing" (Bensimon and Dowd, 2009; Pak et al., 2006) that produces "transfer shock" (Laanan, 2003). In consideration of these challenges, these studies have highlighted a number of prominent solution strategies including summer bridge programs, student cohorts of learning communities, more robust faculty advising, various types of mentoring, and institutional self-studies of transfer to create a more "transfer-amenable" culture (Dowd et al., 2006).

Similarly, researchers have examined the impacts of the STEM curriculum and learning environments on student recruitment and retention in STEM fields, often with a focus on understanding the disproportionate loss of women and underrepresented racial-ethnic minority students from STEM majors at four-year institutions (Aguirre, 2009; Carlone, 2007; Cole and Espinoza, 2008; Crisp, Nora, and Taggart, 2009; Fries-Britt, 1998; Howard-Hamilton et al., 2009; Hurtado et al., 2007, 2009, 2010, 2011; Johnson, 2007; Jones, Barlow, and Villarejo, 2010; McGee and Martin, 2011; Seymour and Hewitt, 1997; Strenta et al., 1994). These case studies characterize concepts such as the culture of science, the notion of a science identity, and the factors that contribute to a sense of belonging or marginalization in STEM classes, majors, and programs. The finding that science and mathematics courses too often function as "weed out" or "gatekeeper" classes that turn students away from STEM majors is a prominent one. Related themes included the emphasis on rigor over instructional support, the demoralizing impact of grading on a curve, stressful and competitive learning environments, and an emphasis on memorization and facts over learning, contextualized problem solving, and application. Enabling students to develop a sense of belonging through faculty and peer interactions in authentic learning activities, particularly research, emerges from these studies as an essential ingredient for STEM reform. The best known and most commonly used innovations that have been developed to address these concerns include various types of active learning and design projects, service learning, bridge programs, learning communities, and other approaches to integrating interdisciplinary curricula (Borrego, Froyd, and Hall, 2010; Henderson, Beach, and Finkelstein, 2011; Vanasupa, Stolk, and Herter, 2009). The need to increase faculty diversity is acknowledged, as is the limited progress in that direction (Stanton-Salazar et al., 2010).

Concerned with the large number of students enrolled in remedial mathematics classes, a special strand of the STEM education literature is focused on developmental education in community colleges. This literature highlights the inadequacies of current practices relative to the scale and complexity of the problem. The prevalence of decontextualized, skills-based instruction and the extended length of the mathematics remedial pathway are often emphasized. Emerging curricular strategies for improved outcomes (in mathematics as well as writing and English language instruction) include placing students in learning communities (Weissman et al., 2011) and implementation of various types of "acceleration" or curricular redesign models, which compress or modularize the curriculum to the essential skills that students need to succeed in their degree-credit courses. Enhanced instructional supports in the form of tutoring, technology-assisted learning, supplemental instruction, intensive advising, and student success courses have also received attention as potential remedies. Another strand of instructional intervention focuses on faculty development through inquiry, data-informed decision making, and institutional self-assessments (Carnegie Foundation for the Advancement of Teaching, 2008). Policy interventions include allowing dual enrollment in high school and college courses and state testing to assess students' college-readiness early enough in their high school years to inform students that they need to increase their level of academic preparedness (Bragg, 2011; Packard, 2011; Rutschow and Schneider, 2011).

A smaller number of studies has specifically examined the transfer experience for STEM students (Malcom, 2008a; Packard et al., 2011; Reyes, 2011; Stanton-Salazar et al., 2010). Bensimon, Dowd, and colleagues examined STEM transfer pathways from community colleges to public universities with the formal designation of Hispanic Serving Institutions (HSIs) (Dowd, Malcom, and Bensimon, 2009; Stanton-Salazar et al., 2010).

Through a case study involving 90 faculty, administrators, and counselors at three universities and three "feeder" community colleges selected as potential exemplars of good practice, they interviewed individuals who had active roles in transfer or STEM transfer programs. The respondents described and shared data showing programs intensively focused on a small number of Hispanic students relative to the entire Hispanic student body at these institutions. As often as not, respondents worked in isolation and were not part of robust networks of faculty and administrators engaged in changing the STEM curriculum. For some, the isolated nature of the work led to a sense that the goal of improving Hispanic student participation and degree completion in STEM fields was not supported by the college leadership. These results highlight the concern that special programs are not adequate to the task of substantially increasing the number of Hispanic students being awarded STEM degrees and the "institutional agents" who work to change the culture of STEM are too few in number to have a systemic impact.

Similarly, Packard (see Appendix B, this volume) emphasizes the importance of faculty mentoring, networks, and advising to encourage women to transfer in STEM fields (see also Packard, 2011). She also found evidence of the value of family and peer academic support. Financial constraints placed stress on many of the 30 women in her sample, two-thirds of whom were first-generation college students and one-quarter were members of racial-ethnic minority groups. As in other studies that have documented the experience of "transfer shock," these transfer students were initially set back by the much quicker pace and rigor of the baccalaureate coursework, especially given that the level of academic support was also lower.

The body of literature focusing specifically on transfer in STEM is not robust enough to substantiate conclusions about the unique programmatic features that are necessary to design effective STEM transfer pathways. However, the intersections in the literature on "choosing and leaving" STEM (Strenta et al., 1994) and the literature on the supports needed for successful transfer suggest that undergraduate research and summer or supplementary bridge programs involving contextualized and active learning are of particular importance. These programs bring students into meaningful relationships with faculty, helping students to develop a science identity and sense of belonging. They also provide a chance for students to see how science is meaningful to their own lives and communities, a factor that is thought to have particular salience for students from underrepresented racial-ethnic minority groups because the STEM faculty and workforce lack role models and mentors with similar backgrounds.

Community College Student Concerns About
the Affordability of Higher Degrees

Community college students are often first-generation students from low-income households (National Center for Education Statistics, 2011, Table 8). Most work either full- or part-time (National Center for Education Statistics, 2011, Table 7). For many, concerns about the affordability of enrolling in a four-year institution cast doubt on the feasibility of transfer (Bensimon and Dowd, 2009; Malcom, 2008a; Ornelas and Solorzano, 2004; Packard et al., 2011). In part, this is due to poor quality financial aid advising and misperceptions of the net price of study at the baccalaureate level once various forms of financial aid are factored in. However, it also reflects a pragmatic outlook and a desire to avoid taking on undergraduate loan debt that they might be unable to pay in the event they do not earn a degree. While students in community colleges and in four-year colleges receive Pell grants (National Center for Education Statistics, 2011, Table 2) and take out loans at similar rates, on average, the amount borrowed by bachelor's degree recipients who started out in public two-year institutions exceeds the amount borrowed by those who started at public four-year institutions (Cataldi et al., 2011, Table 4). In addition, those who start in community colleges have a lower likelihood of earning a bachelor's degree and a higher risk of default (Dowd and Coury, 2006). Data reflecting the period from 2004 to 2009 show that only 19.5 percent of 38 percent of STEM students[2] who begin their studies at public two-year institutions attain a degree or certificate within six years (National Center for Education Statistics, 2011, Table 7). For those who do complete a bachelor's, the time to degree (and opportunity costs for earning and career advancement) is longer (National Center for Education Statistics, 2011, Table 3).

Specifically in regard to STEM, Malcom and Dowd (2012), analyzing NSF's NSRCG data, found that cumulative undergraduate debt among STEM bachelor's degree holders (measured in relative terms in comparison with the typical amount of debt at the graduate's institution) had a negative effect on graduate school enrollment right after college among STEM bachelor's degree holders. Focusing on Hispanic students, they also found that STEM transfer students were more likely to use "self-support" financing strategies, where they used a mix of grants, loans, and earnings, and employer support. This financing profile is consistent with the funding strategies of older, first-generation, and lower-income students who cannot take advantage of parental contributions or loans

---

[2]Includes life sciences, physical sciences, mathematics, computer and information sciences, and engineering and engineering technologies.

(Malcom, Dowd, and Yu, 2010). The available evidence suggests that affordability is a concern for potential STEM transfers, that working off campus may detract from a focus on coursework (National Academy of Sciences, National Academy of Engineering, and Institute of Medicine, 2011), and that aspirations for professional and doctoral degree attainment are dampened due to concerns about debt.

III. PRESTIGE AND THE CULTURE OF SCIENCE

Engineering, the sciences, whether physical, biological or technical, and computing all require mathematical knowledge, reasoning, and skills. These are fields in which epistemic knowledge, which is to say knowledge viewed as objective, rational, and value-neutral, is highly valued (Greenwood and Levin, 2005; Polkinghorne, 2004). Academic disciplines have distinctive cultures and norms (Becher, 1989), in part derived from epistemological paradigms. In academic typologies, STEM fields are considered "hard-pure" (e.g., mathematics and physics) or "hard applied" fields (e.g., engineering), in contrast to "soft-applied" fields (e.g., education and social work) at the other end of the continuum (Austin, 1990). In the hard-pure sciences "knowledge is cumulative and the goals are discovery, explanation, identification of universals, and simplification" (Austin, 1990, p. 64). The hard-applied fields apply such universal knowledge through various forms of engineering, research, and technical design. Despite the fact that engineering, for example, is inherently concerned with social contexts and the public good, these aspects of the engineer's professional responsibilities and identity have become diminished in modern society (Vanasupa, Stolk, and Herter, 2009).

The abstract and generalized truths of hard-pure fields are produced through certain ways of knowing, learning, and thinking, which are called "rational." Success in STEM fields holds prestige in ways that success in other fields does not, because rational knowledge is currently accorded status in U.S. society as an elevated form of knowledge held by experts (Polkinghorne, 2004). Scientists, engineers, and mathematicians, therefore, have a strong identity as rational thinkers. They are also acknowledged survivors or victors who have prevailed in competitive learning environments where producing correct answers and earning high grades are valued. The importance of persistence in the face of repeated error in the inevitable trial and error of scientific research is less clearly acknowledged. Scientific identities are forged in a distinctive "culture of science," with its "gatekeeper" courses and competitive grading (Hurtado et al., 2011). The science culture also promotes ongoing reidentification and association with the scientific community (Austin, 1990; Bergquist and Pawlak, 2008).

It is important to recognize that when scientists, mathematicians, and engineers are asked to invest their professional energies in developing new pedagogies, teaching strategies, and curricula, or to engage in inquiry about the effectiveness of their educational practices, they are being asked to elevate their attention to those aspects of their professional knowledge that are typically accorded less prestige. Education, like social work and counseling, is a soft-applied field, where "knowledge is holistic, and the emphasis is on understanding, interpretation, and particulars" (Austin, 1990, p. 64). In fact, expertise in these fields is defined by one's ability to draw on an extensive repertoire of "particularized" cases and unconsciously select appropriate responses to meet the needs of students or clients. The hallmark of an expert in these fields is the ability to examine an "indeterminate situation," where generalized practices are ineffective in particular cases, and to function effectively under conditions of ambiguity. Educational practice is inherently ambiguous because the teaching-learning relationship is made up of dynamic interactions between teacher and learner (Polkinghorne, 2004)

Without introducing an expectation of adopting reduced academic standards, the Keystone Recommendations of the *Next Generation* report emphasize that the standards of instruction, assessment, and selection into STEM have become too narrow. "Grading on the curve" and the use of "weed out" and "gatekeeper" courses have failed to ensure "opportunities for excellence" or high-quality learning environments for all students across the educational spectrum. Although such practices may be viewed as academically rigorous and necessary by many of those within the STEM professions, researchers have highlighted their negative effect on racial-ethnic minority students and on women (Hurtado et al., 2007, 2009; Seymour and Hewitt, 1997). Subject content and learning environments viewed as value-neutral and objective to some are experienced as "racialized" (Martin, 2009; McGee and Martin, 2011), unsupportive (Lester, 2010), and alienating (Pascarella et al., 1997; Starobin and Laanan, 2008) by others. It may seem paradoxical to individuals steeped and successful in the science culture that the pursuit of scientific knowledge and learning are not neutral and objective activities, experienced in universal terms independent of one's ascribed racial and gender characteristics. Yet, numerous studies (e.g., Howard-Hamilton et al., 2009; Hurtado et al., 2007, 2011; McGee and Martin, 2011) and reports (Institute for Higher Education Policy, n.d.; National Academy of Sciences, National Academy of Engineering, and Institute of Medicine, 2011; Sevo, 2009; Steinecke and Terrell, 2010) provide evidence that students of color and women experience formal STEM postsecondary learning environments as discriminatory, hostile, and alienating. There is now a long history of calls for cultural change in STEM and increased diversity, but the incremental changes have not been sufficient. As observed in the *Crossroads* (National Academy of Sciences, National Academy of Engineering, and Institute of Medicine, 2011) report, the number of African Americans, Hispanics, and Native Americans in certain STEM fields would need to double, triple, or even quadruple to reach parity with the representation of these groups in the U.S. population. Therefore, programs that do not address the fundamental problem of the negative racial climate in STEM fields are unlikely to have a substantial impact to increase diversity.

At this juncture, it is important to note why these considerations are of particular importance when considering strategies to expand STEM transfer pathways between two-year and four-year institutions. First, it is due to the fact that the status differences among fields of study are compounded by the status differences between two-year and four-year college and university faculty. Second, the "chilly climate" of STEM is only harsher for students experiencing the initial "shock" of transfer. Third, students of color are found in community colleges in numbers disproportionately larger than their enrollment in postsecondary education as a whole, which means that efforts to broaden transfer pathways in STEM will have positive equity implications.

The status differences between the two-year and four-year sectors introduce distrust of the quality of the community college curriculum among faculty and administrators who serve on the admissions and curriculum committees of four-year institutions. As a result, the curriculum is poorly aligned and collaboration among faculty is rare (Dowd, 2010; Gabbard et al., 2006; Stanton-Salazar et al., 2010). The negative impact of these poor relationships on students is exacerbated when it comes to transfer in STEM because of the sequential nature of the curriculum.

SECTION IV: EVIDENCE-BASED INNOVATION CONSORTIA

Recent studies of curricular and pedagogical reforms in STEM fields provide evidence that strategies that involve the use of inquiry, reflective practice, and faculty professional development networks are the most promising approaches to bringing about cultural and organizational change (Borrego, Froyd, and Hall, 2010; Henderson, Beach, and Finkelstein, 2011). The dissemination of "best," innovative practices can bring about awareness, but is less effective in leading those on the receiving end of an innovation to the final stage of Rogers' model of diffusion and adoption. These findings are consistent with theories of organizational learning and professional development that emphasize professional knowledge, academic norms, and expertise (Bensimon, 2007; Dowd and Tong, 2007; Kezar and Eckel, 2002; Polkinghorne, 2004; Schein, 1985). They also resonate with models of individual and organizational change, particularly in a situation where professionals are being asked to act as institutional agents to bring about change in their own settings (Seo and Creed, 2002; Stanton-Salazar, 2010). Consequently, recognition of the importance of collective, faculty-based responses to bring about change are growing (Asera, 2008; Kezar, 2012).

Therefore, this report introduces a proposal for the creation of Evidence-Based Inquiry Councils (EBICs), adapted from Dowd and Tong (2007), with a focus on creating effective STEM transfer pathways through the use of inquiry, professional development, and networks. EBICs, as proposed and renamed here as Evidence-Based Innovation Consortia to place the emphasis on *innovation,* would provide an organizational structure to support five institutional roles described in the *Crossroads* report and to foster the "supportive ecosystem" called for in the NSB's *Next Generation* report. To move deliberately in creating STEM learning environments in which a greater number and a more diverse body of students are successful, the *Crossroads* report charged institutions with five roles: leadership, creating a campus-wide commitment to inclusiveness, self-appraisal of the campus climate, plans for constructive change, and ongoing evaluation of implementation efforts.

The EBIC design supports these goals. It also tackles the problem that the transfer structures are not sufficient to support robust transfer pathways in STEM in the absence of interpersonal relationships and shared cultural norms across sectors. Professional development for faculty and college administrators in STEM pedagogy and culturally inclusive practices (Dowd et al., in press) are needed to create such an ecosystem. Such professional development activities will be well received only if they are accorded prestige and provide resources for the production of new knowledge through research, design experiments (Penuel et al., 2011), and inquiry, which is the systematic use of data, reflection, and experimentation to improve professional practices.

The following Keystone Recommendations for the EBIC design are based on those of the *Next Generation* (2010) report:

(1) Keystone Recommendation #1: Provide opportunities for excellence
  (i) Create prestigious research and design centers, called Evidence Based Innovation Consortia, involving STEM faculty in geographic and market-based clusters of two-year and four-year colleges and universities to:
    1. Invent, experiment with, and evaluate innovative approaches to teaching adults foundational mathematics skills and knowledge
    2. Invent, experiment with, and evaluate innovative approaches to active and applied learning
  (ii) Create more intentional mechanisms for diffusion of innovative practices in use in special and supplemental programs to the core curriculum

        (iii) Create a STEM transfer research work-study program through the HEA Reauthorization (for details, see Malcom, 2008a, 2008b) and involve industry in identifying mechanisms to provide work-study positions in collaboration with academic institutions

        (iv) Create public (federal and state) and privately funded STEM transfer scholarships and allocate these to STEM transfer students enrolled in learning communities at the community college and the four-year institution.

(2) Keystone Recommendation #2: Cast a wide net

    (a) Policy Action: Improve talent assessment systems

        (i) Create prestigious research and design centers involving STEM faculty in geographic and market-based clusters of two-year and four-year colleges and universities to develop and validate new forms of diagnostic assessment, student learning assessment, and testing.

    (b) Policy Action: Improve identification of overlooked abilities

        (i) Ensure that students who are successful in special STEM programs find a place in a STEM program and receive necessary mentoring, institutional supports, and opportunities for undergraduate research under the guidance of a faculty member

        (ii) Provide greater investment in the development of a more diverse faculty and administrative workforce in postsecondary education

        (iii) Replace "weed out" and gatekeeper assessments of student learning with talent development assessments

(3) Keystone Recommendation #3: Foster a supportive ecosystem

    (a) Policy Action: Professional development for educators in STEM pedagogy

        (i) Support the development, dissemination, and use of assessment instruments that support deliberate processes of self-appraisal focused on campus climate in STEM learning environments

        (ii) Develop and disseminate models of Culturally Inclusive Pedagogies in STEM

        (iii) Involve STEM educators and educational researchers in joint design and implementation of design experiments, developmental evaluation, and summative evaluation

        (iv) Develop and offer a STEM deans and directors' Leadership Academy and teach participants principles of inquiry and strategies for effective collaboration and institutional self assessment.

        (v) Enroll participants through a three-year membership with staggered terms so that newcomers and experienced members overlap.

REFERENCES

Adelman, C. (2005). *Moving into town—and moving on: The community college in the lives of traditional-age students*. Washington, DC: U.S. Department of Education.

Aguirre, J. (2009). Increasing Latino/a representation in math and science: An insider's look. *Harvard Educational Review, 79*(4), 697-704.

Anderson, G.M., Alfonso, M., and Sun, J.C. (2006). Rethinking cooling out at public community colleges: An examination of fiscal and demographic trends in higher education and the rise of statewide articulation agreements. *Teachers College Record, 108*(3), 422-451.

Anderson, G.M., Sun, J.C., and Alfonso, M. (2006). Effectiveness of statewide articulation agreements on the probability of transfer: A preliminary policy analysis. *Review of Higher Education, 29*(3), 261-291.

Arbona, C., and Nora, A. (2007). The influence of academic and environmental factors on Hispanic college degree attainment. *Review of Higher Education, 30*(3), 247-269.

Asera, R. (2008). *Change and sustain/ability: A program director's reflections on institutional learning*. Stanford, CA: Carnegie Foundation for the Advancement of Teaching.

Attewell, P., Lavin, D., Domina, T., and Levey, T. (2006). New evidence on college remediation. *Journal of Higher Education, 77*(5), 887-924.

Aud, S., Fox, M., and KewalRamani, A. (2010). *Status and trends in the education of racial and ethnic groups*. Washington, DC: U.S. Department of Education, National Center for Education Statistics.

Aud, S., Hussar, W., Kena, G., Blanco, K., Frohlich, L., Kemp, J., et al. (2011). *The condition of education 2011*. Washington, DC: U.S. Department of Education, National Center for Education Statistics.

Austin, A.E. (1990). Faculty cultures, faculty values. In W. G. Tierney (Ed.), *Assessing academic climates and cultures* (vol. 68, pp. 61-74). San Francisco: Jossey-Bass.

Bahr, P.R. (2010). Preparing the underprepared: An analysis of racial disparities in postsecondary mathematics remediation. *Journal of Higher Education, 81*(2), 209-237.

Bailey, T., and Morest, V.S. (Eds.). (2006). *Defending the community college equity agenda*. Baltimore: Johns Hopkins University Press.

Becher, T. (1989). *Academic tribes and territories: Intellectual enquiry and the cultures of disciplines*. Buckingham, UK; Philadelphia, PA: The Society for Research into Higher Education and Open University Press.

Bensimon, E.M. (2007). The underestimated significance of practitioner knowledge in the scholarship of student success. *Review of Higher Education, 30*(4), 441-469.

Bensimon, E.M., and Dowd, A.C. (2009). Dimensions of the "transfer choice" gap: Experiences of Latina and Latino students who navigated transfer pathways. *Harvard Educational Review, 79*(4), 632-658.

Bensimon, E.M., Dowd, A.C., Alford, H., and Trapp, F. (2007). *Missing 87: A study of the "transfer gap" and "choice gap."* Long Beach and Los Angeles: Long Beach City College and the Center for Urban Education, University of Southern California.

Bergquist, W.H., and Pawlak, K. (2008). *Engaging the six cultures of the academy*. San Francisco: Jossey-Bass.

Borrego, M., Froyd, J.E., and Hall, T.S. (2010). Diffusion of engineering education innovations: A survey of awareness and adoption rates in U.S. engineering departments. *Journal of Engineering Education, July*, 185-207.

Bragg, D.D. (2011, December 9). *Two-year college mathematics and student progression in STEM programs of study*. Paper presented at the Community Colleges in the Evolving STEM Education Landscape, Washington, DC.

Brown, R.S., and Niemi, D.N. (2007). *Investigating the alignment of high school and community college assessment in California* (No. 07-3). San Jose, CA: National Center for Public Policy and Higher Education.

Cabrera, A.F., Burkum, K.R., and La Nasa, S.M. (in press). Pathways to a four-year degree: Determinants of transfer and degree completion. In A. Seidman (Ed.), *College student retention: A formula for student success*. Westport, CT: ACE/Prager.

Carlone, H.B. (2007). Understanding the science experiences of successful women of color: Science identity as an analytic lens. *Journal of Research in Science Teaching, 44*(8), 1,187-1,218.

Carnegie Foundation for the Advancement of Teaching. (2008). *Strengthening pre-collegiate education in community colleges: Project summary and recommendations.* A report from Strengthening Pre-collegiate Education in Community Colleges (SPECC). Stanford, CA: Author.

Cataldi, E.F., Green, C., Henke, R., Lew, T., Woo, J., Shepherd, B., et al. (2011). *2008-2009 Baccalaureate and Beyond Longitudinal Study (B&B:08/09): First look.* Washington, DC: U.S. Department of Education, National Center for Education Statistics.

Cejda, B.D. (1998). The effect of academic factors on transfer student persistence and graduation: A community college to liberal arts case study. *Community College Journal of Research and Practice, 22*(7), 675-687.

Cejda, B.D. (2000). Use of the community college in baccalaureate attainment at a private liberal arts college. *Community College Journal of Research and Practice, 24,* 279-288.

Cole, D., and Espinoza, A. (2008). Examining the academic success of Latino students in science technology engineering and mathematics majors. *Journal of College Student Development, 49*(4), 285-300.

Crisp, G., Nora, A., and Taggart, A. (2009). Student characteristics, pre-college, college, and environmental factors as predictors of majoring in and earning a STEM degree: An analysis of students attending a Hispanic-serving institution. *American Educational Research Journal, 46*(4), 924-942.

Dougherty, K.J., and Kienzl, G.S. (2006). It's not enough to get through the open door: Inequalities by social background in transfer from community colleges to four-year colleges. *Teachers College Record, 108*(3), 452-487.

Dowd, A.C. (2008). The community college as gateway and gatekeeper: Moving beyond the access "saga" to outcome equity. *Harvard Educational Review, 77*(4), 407-419.

Dowd, A.C. (2010). Improving transfer access for low-income community college students. In A. Kezar (Ed.), *Recognizing and serving low-income students in postsecondary education: An examination of institutional policies, practices, and culture* (pp. 217-231). New York: Routledge.

Dowd, A.C., and Coury, T. (2006). The effect of loans on the persistence and attainment of community college students. *Research in Higher Education, 47*(1), 33-62.

Dowd, A.C., and Melguizo, T. (2008). Socioeconomic stratification of community college transfer access in the 1980s and 1990s: Evidence from HS&B and NELS. *Review of Higher Education, 31*(4), 377-400.

Dowd, A.C., and Tong, V.P. (2007). Accountability, assessment, and the scholarship of "best practice." In J.C. Smart (Ed.), *Handbook of higher education* (vol. 22, pp. 57-119). New York: Springer.

Dowd, A.C., Bensimon, E.M., Gabbard, G., Singleton, S., Macias, E.E., Dee, J., et al. (2006). *Transfer access to elite colleges and universities in the United States: Threading the needle of the American dream.* Lansdowne, VA: Jack Kent Cooke Foundation.

Dowd, A.C., Cheslock, J.J., and Melguizo, T. (2008). Transfer access from community colleges and the distribution of elite higher education. *Journal of Higher Education, 79*(4), 442-472.

Dowd, A.C., Malcom, L.E., and Bensimon, E.M. (2009). *Benchmarking the success of Latina and Latino students in STEM to achieve national graduation goals.* Los Angeles: Center for Urban Education, University of Southern California.

Dowd, A.C., Malcom, L.E., and Macias, E.E. (2010). *Improving transfer access to STEM bachelor's degrees at Hispanic-serving institutions through the America COMPETES Act.* Los Angeles: Center for Urban Education, University of Southern California.

Dowd, A.C., Sawatzky, M., Rall, R.M., and Bensimon, E.M. (in press). Action research: An essential practice for Twenty-First Century assessment. In R.T. Palmer, D.C. Maramba, and M. Gasman (Eds.), *Fostering success of ethnic and racial minorities in STEM: The role of minority-serving institutions.* New York: Routledge.

Fries-Britt, S. (1998). Moving beyond black achiever isolation: Experiences of gifted black collegians. *The Journal of Higher Education, 69*(5), 556-576.

Gabbard, G., Singleton, S., Macias, E.E., Dee, J., Bensimon, E.M., Dowd, A.C., et al. (2006). *Practices supporting transfer of low-income community college transfer students to selective institutions: Case study findings.* Boston and Los Angeles: University of Massachusetts and University of Southern California.

Greenwood, D.J., and Levin, M. (2005). Reform of the social sciences and of universities through action research. In N.K. Denzin and Y.S. Lincoln (Eds.), *Handbook of qualitative research* (3rd ed., pp. 43-64). Thousand Oaks, CA: Sage.

Grubb, W.N., Boner, E., Frankel, K., Parker, L., Patterson, D., Gabriner, R., et al. (2011). *Understanding the "crisis" in basic skills: Framing the issue in community colleges.* Berkeley: Policy Analysis for California Education.

Hagedorn, L.S., and DuBray, D. (2010). Math and science success and nonsuccess: Journeys within the community college. *Journal of Women and Minorities in Science and Engineering, 16*(1), 31-50.

Hardy, D., and Katsinas, S.G. (2010). Changing STEM associate's degree production in public associates' colleges 1990 to 2005: Exploring institutional type, gender, and field of study. *Journal of Women and Minorities in Science and Engineering, 16*(1), 7-30.

Henderson, C., Beach, A., and Finkelstein, N. (2011). Facilitating change in undergraduate STEM instructional practices: An analytic review of the literature. *Journal of Research in Science Teaching, 48*(8), 952-984.

Hoachlander, G., Sidora, A., and Horn, L. (2003). *Community college students: Goals, academic preparation, and outcomes* (Postsecondary Education Descriptive Analysis Reports No. 2003-164). Washington, DC: National Center for Education Statistics.

Horn, L. (2006). *Placing college graduation rates in context: How 4-year college graduation rates vary with selectivity and the size of low-income enrollment* (No. 2006-184). Washington, DC.: U.S. Department of Education. National Center for Education Statistics.

Horn, L., and Lew, S. (n.d.). *California community college transfer rates: Who is counted makes a difference.* MPR Research Brief #1.

Howard-Hamilton, M.F., Morelon-Quainoo, C.L., Johnson, S.D., Winkle-Wagner, R., and Santiague, L. (Eds.). (2009). *Standing on the outside looking in: Underrepresented students' experiences in advanced-degree programs.* Sterling, VA: Stylus.

Hughes, K.L., and Scott-Clayton, J. (2011). *Assessing developmental assessment in community colleges.* New York: Community College Research Center, Teachers College, Columbia University.

Hurtado, S., Han, J.C., Sáenz, V.B., Espinosa, L.L., Cabrera, N.L., and Cerna, O.S. (2007). Predicting transition and adjustment to college: Minority biomedical and behavioral sciences. *Research in Higher Education, 48*(7), 841-887.

Hurtado, S., Cabrera, N.L., Lin, M.H., Arellano, L., and Espinosa, L.L. (2009). Diversifying science: Underrepresented student experiences in structured research program. *Research in Higher Education, 50*(2), 189-214.

Hurtado, S., Pryor, J., Tran, S., Blake, L.P., DeAngelo, L., and Aragon, M. (2010). *Degrees of success: Bachelor's degree completion rates among initial STEM majors.* Los Angeles: Higher Education Research Institute, University of California.

Hurtado, S., Eagan, M.K., Tran, M.C., Newman, C.B., Chang, M.J., and Velasco, P. (2011). "We do science here": Underrepresented students' interaction with faculty in different college contexts. *Journal of Social Issues, 67*(3), 553-579.

Institute for Higher Education Policy. (n.d.). *Diversifying the STEM pipeline: The Model Replication Institutions Program*. Washington, DC: Author.

Johnson, A.C. (2007). Unintended consequences: How science professors discourage women of color. *Science Education, 91*(5), 805-821.

Jones, M.T., Barlow, A.E.L., and Villarejo, M. (2010). Importance of undergraduate research for minority persistence and achievement in biology. *Journal of Higher Education, 81*(1), 82-115.

Kezar, A. (2012). The path to pedagogical reform in the sciences. *Liberal Education,* 40-45.

Kezar, A., and Eckel, P.D. (2002). The effect of institutional culture on change strategies in higher education: Universal principles or culturally responsive concepts? *Journal of Higher Education, 73*(4), 436-460.

Kienzl, G.S., Wesaw, A.J., and Kumar, A. (2011). *Understanding the transfer process*. Washington, DC: Institute for Higher Education Policy.

Kirst, M. (2007, Winter). Who needs it?: Identifying the proportion of students who require postsecondary remedial education is virtually impossible. *National CrossTalk, 15,* 11-12.

Laanan, F.S. (1996). Making the transition: Understanding the adjustment process of community college transfer students. *Community College Review, 23*(4), 69-84.

Laanan, F.S. (2003). Degree aspirations of two-year college students. *Community College Journal of Research and Practice, 27,* 495-518.

Lee, J.M., and Ranson, T. (2011). *The educational experience of young men of color: A review of research, pathways, and progress*. New York: The College Board.

Lester, J. (2010). Women in male-dominated career and technical education programs at community colleges: Barriers to participation and success. *Journal of Women and Minorities in Science and Engineering, 16*(1), 51-66.

Levin, H.M., and Calcagno, J.C. (2008). Remediation in the community college: An evaluator's perspective. CCRC Working Paper. *Community College Review, 35,* 181-207.

Malcom, L.E. (2008a). *Accumulating (dis)advantage? Institutional and financial aid pathways of Latino STEM baccalaureates*. Unpublished dissertation, University of Southern California, Los Angeles.

Malcom, L.E. (2008b). *Multiple pathways to STEM: Examining state differences in community college attendance among Latino STEM bachelor's degree holders*. Presented at the Association for the Study of Higher Education, November, Jacksonville, FL.

Malcom, L.E., and Dowd, A.C. (2012). The impact of undergraduate debt on the graduate school enrollment of STEM baccalaureates. *Review of Higher Education, 35*(2), 265-305.

Malcom, L.E., Dowd, A.C., and Yu, T. (2010). *Tapping HSI-STEM funds to improve Latina and Latino access to STEM professions*. Los Angeles: Center for Urban Education, University of Southern California.

Martin, D.B. (2009). Researching race in mathematics. *Teachers College Record, 111*(2), 295-338.

McGee, E., and Martin, D.B. (2011). "You would not believe what I have to go through to prove my intellectual value!" Stereotype management among academically successful black mathematics and engineering students. *American Educational Research Journal, 48*(6), 1,347-1,389.

Melguizo, T., and Dowd, A.C. (2009). Baccalaureate success of transfers and rising four-year college juniors. *Teachers College Record, 111*(1), 55-89.

Melguizo, T., Hagedorn, L.S., and Cypers, S. (2008). Remedial/developmental education and the cost of community college transfer: A Los Angeles County sample. *Review of Higher Education, 31*(4), 401-431.

Mullin, C.M. (2011). *The road ahead: A look at trends in the educational attainment of community college students*. Washington, DC: American Association of Community Colleges.

National Academy of Sciences, National Academy of Engineering, and Institute of Medicine. (2011). *Expanding underrepresented minority participation: America's science and technology talent at the crossroads.* Committee on Underrepresented Groups and the Expansion of the Science and Engineering Workforce Pipeline; Committee on Science, Engineering, and Public Policy; Policy and Global Affairs. Washington, DC: The National Academies Press.

National Center for Education Statistics. (2011). *Students attending for-profit postsecondary institutions: Demographics, enrollment characteristics, and 6-year outcomes.* Washington, DC: U.S. Department of Education, National Center for Education Statistics, Institute of Education Sciences.

National Center for Education Statistics. (n.d.-a). *Table A-40-1: Number of associate's and bachelor's degrees awarded by degree-granting institutions, percentage of total, number and percentage awarded to females, and percent change, by selected fields of study: Academic years 1998-99 and 2008-09.* Available: http://nces.ed.gov/surveys/ctes/tables/index.asp?LEVEL=COLLEGE [January 25, 2012].

National Center for Education Statistics. (n.d.-b). *Table P45: Percentage distribution of credential-seeking undergraduates, by sex, race/ethnicity, age, credential goal, and curriculum area: 2007-08.* Available: http://nces.ed.gov/surveys/ctes/tables/index.asp?LEVEL=COLLEGE [January 25, 2012].

National Center for Education Statistics. (n.d.-c). *Table P46: Percentage distribution of credential-seeking undergraduates in career education, by sex, race/ethnicity, age, credential goal, and career field of study: 2007-08.* Available: http://nces.ed.gov/surveys/ctes/tables/index.asp?LEVEL=COLLEGE [January 25, 2012].

National Science Board. (2010). *Preparing the next generation of STEM innovators: Identifying and developing our national human capital.* Arlington, VA: Author.

National Science Foundation. (2011). *Women, minorities, and persons with disabilities in science and engineering: 2011* (No. NSF 04-317.) Arlington, VA: Author.

Ornelas, A., and Solorzano, D.G. (2004). Transfer conditions of Latina/o community college students: A single institution case study. *Community College Journal of Research and Practice, 28*(3), 233-248.

Packard, B.W.-L. (2011). *Effective outreach, recruitment, and mentoring into STEM pathways: Strengthening partnerships with community colleges.* Paper presented at the Community Colleges in the Evolving STEM Education Landscape, Washington, DC.

Packard, B.W.-L., Gagnon, J.L., LaBelle, O., Jeffers, K., and Lynn, E. (2011). **Women's experiences in the STEM community college transfer pathway.** *Journal of Women and Minorities in Science and Engineering, 17*(2), 129-147.

Pak, J., Bensimon, E.M., Malcom, L.E., Marquez, A., and Park, D. (2006). *The life histories of ten individuals who crossed the border between community colleges and selective four-year colleges.* Los Angeles: University of Southern California.

Parsad, B., Lewis, L., and Greene, B. (2003). *Remedial education at degree-granting postsecondary institutions in Fall 2000.* (No. NCES 2004-010.) Washington, DC: U.S. Department of Education, National Center for Education Statistics.

Pascarella, E.T., Whitt, E.J., Edison, M.I., Nora, A., Hagedorn, L.S., Yeager, P.M., et al. (1997). Women's perceptions of a "chilly climate" and their cognitive outcomes during the first year of college. *Journal of College Student Development, 38*(2), 109-124.

Penuel, W.R., Fishman, B.J., Cheng, B.H., and Sabelli, N. (2011). Organizing research and development at the intersection of learning, implementation, and design. *Educational Researcher, 40*(7), 331-337.

Polkinghorne, D.E. (2004). *Practice and the human sciences: The case for a judgment-based practice of care.* Albany: State University of New York Press.

Reyes, M.-E. (2011). Unique challenges for women of color in STEM transferring from community college to universities. *Harvard Educational Review, 81*(2), 241-263.

Rutschow, E.Z., and Schneider, E. (2011). *Unlocking the gate: What we know about improving developmental evaluation.* New York: MDRC.

Ruud, C.M., and Bragg, D.D. (2011). *The applied baccalaureate: What we know, what we learned, and what we need to know.* Champaign: Office of Community College Research and Leadership, University of Illinois at Urbana–Champaign.

Santiago, D. (2006). *Inventing Hispanic-Serving Institutions (HSIs): The basics.* Washington, DC: Excelencia in Education.

Schein, E.H. (1985). Understanding culture change in the context of organizational change. *Organizational Culture and Leadership,* 244-310.

Seo, M.G., and Creed, W.E.D. (2002). Institutional contradictions, praxis, and institutional change: A dialectical perspective. *Academy of Management Review, 27*(2), 222-247.

Sevo, R. (2009). *The talent crisis in science and engineering.* Available: http://www.engr.psu.edu/AWE/ARPResources.aspx [February 1, 2009].

Seymour, E., and Hewitt, N.C. (1997). *Talking about leaving: Why undergraduates leave the sciences.* Boulder, CO: Westview Press.

Snyder, T.D., Tan, A.G., and Hoffman, C.M. (2006). *Digest of education statistics, 2005.* Washington, DC: National Center for Education Statistics.

Stanton-Salazar, R.D. (2010). A social capital framework for the study of institutional agents and their role in the empowerment of low-status youth. *Youth and Society, 42*(2), 1-44.

Stanton-Salazar, R.D., Macias, R.M., Bensimon, E.M., and Dowd, A.C. (2010). *The role of institutional agents in providing institutional support to Latino students in STEM.* Paper presented at the Association for the Study of Higher Education.

Starobin, S., and Laanan, F.S. (2008). Broadening female participation in science, technology, engineering, and mathematics: Experiences at community colleges. In J. Leaster (Ed.), *Gendered perspectives on community college* (pp. 37-46). Wilmington, DE: Wiley Periodicals.

Steinecke, A., and Terrell, C. (2010). Progress for the future? The impact of the Flexner Report on medical education for racial and ethnic minority physicians in the United States. *Academic Medicine, 85*(2), 236-245.

Strenta, A.C., Elliott, R., Adair, R., Matier, M., and Scott, J.W. (1994). Choosing and leaving science in highly selective institutions. *Research in Higher Education, 35*(5), 513-547.

Townsend, B.K., and Wilson, K. (2006). "A hand hold for a little bit": Factors facilitating the success of community college transfer students to a large research university. *Journal of College Student Development, 47*(4), 439.

Vanasupa, L., Stolk, J., and Herter, R.J. (2009). The four-domain development diagram: A guide for holistic design of effective learning experiences for the twenty-first century engineer. *Journal of Engineering Education, 98*(1), 667-680.

Wang, X. (2011). *Modeling student entrance into STEM fields of study at community colleges and four-year institutions: Towards a theoretical framework of motivation, high school learning, and postsecondary context of support.* Proposal submitted to Association for Institutional Research.

Weissman, E., Butcher, K., Schneider, E., Teres, J.J., Collado, H., and Greenberg, D. (2011). *Learning communities for students in developmental math: Impact studies at Queensborough and Houston Community Colleges.* New York: National Center for Postsecondary Research.

Zinser, R., and Hanssen, C. (2006). Improving access to the baccalaureate: Articulation agreements and the National Science Foundation's Advanced Technological Education program. *Community College Review 34*(1), 27-43.

Appendix E

Brief Biographies of Committee Members and Staff

COMMITTEE

GEORGE R. BOGGS *(Chair)* is president and chief executive officer emeritus of the American Association of Community Colleges where he served for 10 years, and is superintendent and president emeritus of Palomar College, where he served for 15 years. He previously served as faculty member, division chair, and associate dean of instruction at Butte College in California. He is currently a member of the Board on Science Education and was one of the original members of the former Committee on Undergraduate Science Education. He has served on several National Science Foundation panels and committees. He holds a bachelor's degree in chemistry from Ohio State University, a master's degree in chemistry from the University of California at Santa Barbara, and a Ph.D. in educational administration from the University of Texas at Austin.

THOMAS R. BAILEY is the George and Abby O'Neill professor of economics and education in the Department of International and Transcultural Studies at Teachers College, Columbia University. In 1996, with support from the Alfred P. Sloan Foundation, he established the Community College Research Center at Teachers College, which conducts a large portfolio of qualitative and quantitative research based on fieldwork at community colleges and analysis of national- and state-level datasets. In July 2006, he became the director of the National Center for Postsecondary Research), funded by a grant from the Institute of Education Sciences of the U.S. Department of Education. Since 1992, he has also been the director of the Institute on Education and the Economy at Teachers College. His most recent book, co-edited with Vanessa Morest, is *Defending the Community College Equity Agenda* (Johns Hopkins University Press, 2006). He is an economist, with specialties in education, labor economics, and econometrics, and holds a Ph.D. in labor economics from MIT.

LINNEA FLETCHER is chair of the Department of Biotechnology at Austin Community College (ACC) and continues to work on several federal and state grants. After several years working in laboratories, she began a career in education at ACC in 1991. She served as chair of the Department of Biology and then assistant dean of mathematics, science and technology. In 1997, she became interested in bringing biotechnology training to ACC. As part of this goal, she joined the Advanced Technological Education Center of Excellence for Biotechnology and Life Sciences Grant (Bio-Link). After serving several years as chair of the Department of Biotechnology and working on a variety of National Science Foundation (NSF) educational grants, she accepted a rotator position at NSF as a program director in the Division of Undergraduate Education. After two years at NSF, she returned to ACC. She obtained her bachelor's and master's degrees in biology, chemistry, and biochemistry at the University of California at Irvine and her Ph.D. in microbiology at the University of Texas at Austin.

BRIDGET TERRY LONG is the Xander Professor of Education and Economics at Harvard University, Graduate School of Education. She is a faculty research associate of the National Bureau of Economic Research and research affiliate of the National Center for Postsecondary Research. In July 2005, the *Chronicle of Higher Education* featured her as one of the "New Voices" in higher education; and in 2008, the National Association of Student Financial Aid Administrators awarded her the Robert P. Huff Golden Quill Award for excellence in research and published works on student financial assistance. An economist specializing in education, she studies the transition from high school to higher education and beyond. Her work focuses on college access and choice, factors that influence student outcomes, and the behavior of postsecondary institutions. She received an A.B. from Princeton University and M.A. and Ph.D. from the Harvard University Department of Economics.

JUDY C. MINER is president of Foothill College in Los Altos Hills, California. She has worked as a higher education administrator since 1977 and in the California Community Colleges since 1979 where she has held administrative positions in student services and instruction at City College of San Francisco, the California Community Colleges Chancellor's

Office and De Anza College. She serves on the Commission on Inclusion of the American Council of Education, the Board of Directors for the Council of Higher Education Accreditation, and the STEM Higher Education Working Group under the auspices of the President's Council of Advisors on Science and Technology. She earned her B.A. in history and French and M.A. in history from Lone Mountain College in San Francisco and her Ed.D. in organization and leadership (with a concentration in education law) from the University of San Francisco.

KARL S. PISTER, chair of the governing board of the California Council on Science and Technology, is former vice president—educational outreach, of the University of California, and chancellor emeritus of the University of California, Santa Cruz. Prior to retirement, he completed five decades of service to higher education, beginning his career as assistant professor in the Department of Civil Engineering at University of California, Berkeley. He served as chairman of the Division of Structural Engineering and Structural Mechanics before his appointment as dean of the College of Engineering in 1980, a position he held for 10 years. From 1985 to 1990, he was the first holder of the Roy W. Carlson chair in engineering. From 1991 to 1996, he served as chancellor, University of California, Santa Cruz. He is a member of the National Academy of Engineering and a fellow of the American Academy of Arts and Sciences. He is also a fellow of the American Academy of Mechanics, American Society of Mechanical Engineers, and American Association for the Advancement of Science, and an honorary fellow of the California Academy of Sciences. He is a member of the Board of Directors of the Monterey Bay Aquarium Research Institute, Center for the Future of Teaching and Learning, and Board of Trustees of the American University of Armenia. He also served as founding chairman of the Board on Engineering Education of the National Research Council. He has a Ph.D. in theoretical and applied mechanics from the University of Illinois at Urbana–Champaign.

NATIONAL RESEARCH COUNCIL STAFF

JAY B. LABOV (*PI*) is senior advisor for education and communication, National Academy of Sciences and the National Research Council (NRC), and serves as the principal investigator (PI) for this project. He has worked for the Academies for 15 years. He also serves as program director for Biology Education with the NRC's Board on Life Sciences and the director of the National Academies Teacher Advisory Council. In this role, he is responsible for coordinating all aspects of the work for this project with the co-PIs and with program, administrative, and financial management staff at the National Academies. He also works with the

Board on Life Sciences and the Teacher Advisory Council to obtain their input and serves as the National Academies' representative and liaison with staff in the Division of Undergraduate Education and the Advanced Technological Education program of the National Science Foundation. He earned his B.S. in biology from the University of Miami, and M.S. in zoology and Ph.D. in biological sciences from the University of Rhode Island.

CATHERINE DIDION (*co-PI*) is senior program officer, National Academy of Engineering (NAE). Her portfolio includes the Diversity of the Engineering Workforce Program with a charge to provide staff leadership to the NAE's efforts to enhance the diversity of the engineering workforce at all levels including the diversity of those being prepared to enter the future workforce. She is the project director of a $2 million National Science Foundation (NSF) grant to increase the number of women receiving baccalaureate degrees in engineering and helped organize a workshop in 2010 for NSF on the underrepresentation of minority males in STEM. As a co-PI on the project, Didion links the work of the NAE on two-year and four-year engineering articulation agreements that should facilitate greater participation of community college students in engineering education. She also engages the relevant engineering societies and associations that are working with two-year institutions. She received an A.B. in international relations from Mount Holyoke College and a graduate certificate of achievement in e-commerce from the University of Virginia.

PETER H. HENDERSON (*co-PI*) is director of the NRC's Board on Higher Education and Workforce (BHEW). His areas of specialization include higher education policy, labor markets for scientists and engineers, and federal science and technology research funding. He currently directs BHEW's Study on Research Universities and the Committee on Science, Engineering, and Public Policy's study on underrepresented groups and the expansion of the science and engineering workforce pipeline, both of which follow from the National Academies' *Rising Above the Gathering Storm*. He has previously contributed as study director or staff to a variety of NRC education and workforce studies germane to this project, including *Science Professionals: Master's Education for a Competitive World* (2009); *Enhancing the Community College Pathway to Engineering Careers* (2005); and *Building a Workforce for the Information Economy* (2001). He received his B.A. from Johns Hopkins University, master's in public policy from Harvard University's John F. Kennedy School of Government, and Ph.D. in history from Johns Hopkins University.

MARGARET L. HILTON, is a senior program officer with the Board on Science Education and has led a series of National Research Council activities exploring emerging workforce skill demands and K-12 and higher education to meet those demands. She has convened a planning meeting on career-technical education and workshops on future skill demands, promising practices in undergraduate science, technology, engineering, and mathematics, and the intersection of science education and 21st century skills. She has directed studies of high school science laboratories, the Occupational Information Network (O*NET) database and the use of simulations and games for science learning. She is currently directing a study on Defining Deeper Learning and 21st Century Skills. She has a B.A. in geography from the University of Michigan, a master of regional planning from the University of North Carolina at Chapel Hill, and master of arts in education and human development from George Washington University.

MARY ANN KASPER is a senior program assistant in the Division of Behavioral and Social Sciences and Education at the National Research Council. She has assisted on many projects and their reports, among them: *America's Lab Report: Investigations in High School Science*, *International Education and Foreign Languages: Keys to Securing America's Future*, *Mathematics Learning in Early Childhood: Paths Toward Excellence and Equity*, and *Student Mobility: Exploring the Impacts of Frequent Moves on Achievement*. She is presently assisting with other projects, including a study on adolescent and adult literacy, and a report on the use of social science evidence for public policy.

MARTIN STORKSDIECK (*co-PI*) serves as director of the Board on Science Education (BOSE) where he oversees studies that address a wide range of issues with connections to this project (e.g., climate change education, developing a conceptual framework for new science education standards, and discipline-based education research in higher education). As a co-PI, he will link the work of the ad-hoc committee to relevant work being undertaken by BOSE, especially the development of a proposed study on Barriers and Opportunities in Completing Two- and Four-Year STEM Degrees. He holds an M.S. in biology from Albert-Ludwigs University in Freiburg, Germany, an M.P.A. from Harvard University's Kennedy School of Government, and a Ph.D. in education from Leuphana University in Lüneburg, Germany.

CYNTHIA WEI was a National Academies Christine Mirzayan Science & Technology Policy Fellow working with Dr. Labov at the time of the summit. She recently completed a AAAS Science & Technology Policy Fellowship at the National Science Foundation in the Division of Undergraduate Education, where she worked on a wide range of issues in science, technology, engineering, and mathematics education with a primary focus on biology education and climate change education. She has diverse teaching experiences as a K-6 general science teacher, high school biology teacher, and college-level biology instructor. She received a dual-degree PhD in zoology and ecology, evolutionary biology, and behavior from Michigan State University, and a B.A. in neurobiology and behavior from Cornell University. She also was a postdoctoral research associate at the University of Nebraska, Lincoln's Avian Cognition Laboratory.

Appendix F

Brief Biographies of
Presenters and Panelists

ERIC BETTINGER is an associate professor in the Education and the Economics departments at Stanford University. Prior to joining the faculty at Stanford, he was an associate professor of economics at Case Western Reserve University. He has done wide-reaching research on how organizational structure and policy influence educational achievement of students of different race, gender, and income. He is also studying what factors determine student success in college. His work aims to bring understanding of these cause-and-effect relationships in higher education. He earned his B.A. from Brigham Young University and Ph.D. from Massachusetts Institute of Technology.

GEORGE BOGGS: See biographical sketch in Appendix E.

DEBRA D. BRAGG is a professor in the Department of Education Organization, Policy and Leadership at the University of Illinois. She is also director of the Office of Community College Research and Leadership and director of the Forum on the Future of Public Education, a strategic initiative of the College of Education at Illinois. Her research focuses on P-20 policy, with a special interest in the transition of youth and adults to college. She has led research and evaluation studies funded by federal, state, and foundation sponsors, including examining the participation of underserved students in college transition and career pathways. Recent studies include evaluations of bridge-to-college programs funded by the Joyce Foundation and the U.S. Department of Education and applied baccalaureate programs funded by the Lumina Foundation for Education. She is the recipient of the career teaching and distinguished research awards from the College of Education at the University of Illinois, and the senior scholar award from the Council for the Study of Community Colleges. She holds a bachelor's degree from the University of Illinois and master's and Ph.D. degrees from Ohio State University.

V. CELESTE CARTER is program director of the Division of Undergraduate Education (DUE) at the National Science Foundation (NSF). She joined the Division of Biological and Health Sciences at Foothill College in 1994 to develop and head a biotechnology program. She served as a DUE program director twice as a rotator and accepted a permanent program director position in 2009. She is the lead program director for the Advanced Technological Education (ATE) Program in DUE, as well as working on other programs in the division and across NSF. She received her Ph.D. in microbiology from the Pennsylvania State University School of Medicine in 1982 and completed postdoctoral studies in the laboratory of Dr. G. Steven Martin at the University of California, Berkeley.

ALICIA C. DOWD is an associate professor of higher education at the University of Southern California's Rossier School of Education and co-director of the Center for Urban Education (CUE). Her research focuses on political-economic issues of racial-ethnic equity in postsecondary outcomes, organizational learning and effectiveness, accountability, and the factors affecting student attainment in higher education. She is the principal investigator of the National Science Foundation-funded study *Pathways to STEM Bachelor's and Graduate Degrees for Hispanic Students and the Role of Hispanic Serving Institutions*. As a research methodologist, she has also served on numerous federal evaluation and review panels. She was awarded a B.A. in English literature at Cornell University and a doctorate at Cornell, where she studied the economics and social foundations of education, labor economics, and curriculum and instruction.

HARVEY V. FINEBERG is president of the Institute of Medicine (IOM). He previously served Harvard University as provost for four years and 13 years as dean of the School of Public Health. He helped found and served as president of the Society for Medical Decision Making and has been a consultant to the World Health Organization. His research has included assessment of medical technology, evaluation of vaccines, and dissemination of medical innovations. At IOM, he has chaired and served on a number of panels dealing with health policy issues, ranging from AIDS to new medical technology. He also served as a member of the Public

Health Council of Massachusetts (1976-1979), as chairman of the Health Care Technology Study Section of the National Center for Health Services Research (1982-1985), and as president of the Association of Schools of Public Health (1995-1996). He is the author or co-author of numerous books and articles on subjects ranging from AIDS prevention to medical education. He holds four degrees from Harvard, including M.D. and Ph.D. in public policy.

TOBY HORN is co-director of the Carnegie Academy for Science Education at the Carnegie Institution of Washington. She joined the faculty of Thomas Jefferson High School for Science and Technology in Fairfax County, VA, two weeks before the doors opened in 1985. As co-director of the Life Science and Biotechnology Laboratory for nearly 14 years, she developed one of the first high school biotechnology programs for students in grades 9-12. After two years as outreach coordinator for the Fralin Biotechnology Center at Virginia Tech, she joined the Carnegie Institution to work in District of Columbia Public Schools. Other relevant activities include membership on the NAS committee to revise Science, Evolution, and Creationism, recipient of the Bruce Alberts Award (2009), president of the National Association of Biology Teachers (2006), and current member on the National Visiting Committee for the Bio-Link National Center of Excellence. She holds an A.B. in chemistry from Bryn Mawr College and a Ph.D. in MCD biology from the University of Colorado, Boulder. She was also a staff fellow at the National Cancer Institute for five years.

FREEMAN A. HRABOWSKI, III, has served as president of the University of Maryland, Baltimore County (UMBC) since 1992. His research and publications focus on science and math education, with special emphasis on minority participation and performance. He chaired the National Academies committee that recently produced the report *Expanding Underrepresented Minority Participation: America's Science and Technology Talent at the Crossroads*. In 2008, he was named one of America's Best Leaders by *U.S. News & World Report*, which in 2009, 2010, and 2011 ranked UMBC as the #1 "Up and Coming" university in the nation. In 2009, *Time* magazine named him one of America's 10 Best College Presidents. In 2011, he received the TIAA-CREF Theodore M. Hesburgh Award for Leadership Excellence and the Carnegie Corporation of New York's Academic Leadership Award. With philanthropist Robert Meyerhoff, he co-founded the Meyerhoff Scholars Program in 1988, considered a national model. He has authored numerous articles and co-authored two books, *Beating the Odds* and *Overcoming the Odds* (Oxford University Press), focusing on parenting and high-achieving African American males and females in science.

MARTHA J. KANTER was nominated by President Barack Obama on April 29, 2009, to be the under secretary of education and was confirmed by the Senate on June 19, 2009. She oversees policies, programs, and activities related to postsecondary education, adult and career-technical education, federal student aid, and five White House Initiatives. She is the first community college leader to serve in the under secretary position. From 2003 to 2009, she served as chancellor of the Foothill-De Anza Community College District, one of the largest community college districts in the nation. In 1977, after serving as an alternative high school teacher in Massachusetts and New York, she established the first program for students with learning disabilities at San Jose City College. She then served as a director, dean, and vice chancellor for policy and research for the California Community Colleges Chancellor's Office in Sacramento. In 1990, she returned to San Jose City College as vice president of instruction and student services until she was named president of De Anza College in 1993. She received a bachelor's degree in sociology from Brandeis University, master's degree in education with a concentration in clinical psychology and public practice from Harvard University, and a doctorate in organization and leadership from the University of San Francisco. She also holds honorary degrees from Palo Alto University, Chatham University, Lakes Region Community College, Moraine Valley Community College, and the Alamo Colleges.

JAY B. LABOV: See biographical sketch in Appendix E.

JANE OATES was nominated by President Barack Obama in April 8, 2009, and confirmed as assistant secretary for employment and training on June 19, 2009. She leads the Employment and Training Administration in its mission to design and deliver high-quality training and employment programs. Prior to her appointment, she served as executive director of the New Jersey Commission on Higher Education and senior advisor to Governor Jon S. Corzine. She served for nearly a decade as senior policy advisor for Massachusetts Senator Edward M. Kennedy. She began her career as a teacher in the Boston and Philadelphia public schools and later as a field researcher at Temple University's Center for Research in Human Development and Education. She received her B.A. in education from Boston College and M.Ed. in reading from Arcadia University.

BARBARA M. OLDS is acting deputy assistant director and senior advisor to the Directorate for Education and Human Resources of the National Science Foundation, where she focuses on issues related to international science and engineering education, program and project evaluation, and education and education research policy. She previously served in the directorate as an expert/consultant on education issues, as division director for the Division on Research, Evaluation and Communication, and as acting division director for the Division of Elementary, Secondary, and Informal Education. She is professor emerita of liberal arts and international studies at the Colorado School of Mines. During her long career there, she served as the director of the Engineering Practices Introductory Course Sequence, as the director of the McBride Honors Program in Public Affairs for Engineers, and as the associate provost for educational innovation. Her research interests lie primarily in understanding and assessing engineering student learning. She has participated in a number of curriculum innovation projects and has been active in the engineering education research and evaluation communities. She is a fellow of the American Society for Engineering Education, a senior editor for the *Journal of Engineering Education*, and was a Fulbright lecturer/researcher in Sweden. She holds an undergraduate degree from Stanford University, and an M.A. and Ph.D. from the University of Denver, all in English.

BECKY WAI-LING PACKARD is a professor of educational psychology at Mount Holyoke College. She is also the co-director of the Weissman Center for Leadership and the Liberal Arts with responsibility for teaching and faculty development initiatives. Her research, funded by the National Science Foundation, focuses on the mentoring and persistence of students from first-generation for college and lower-income backgrounds as they navigate trade, work, community college transfer, and four-year college pathways in science and engineering fields. In 2005, she received the Presidential Early Career Award for Science and Engineering at the White House. She has published numerous articles on this topic and frequently works with colleges, community organizations, and businesses to design formalized mentoring programs and effective advising practices. She earned her B.A. from the University of Michigan and Ph.D. in educational psychology from Michigan State University.